Alfred Earl

The living organism

An introduction to the problems of biology

Alfred Earl

The living organism
An introduction to the problems of biology

ISBN/EAN: 9783337216764

Printed in Europe, USA, Canada, Australia, Japan

Cover: Foto ©berggeist007 / pixelio.de

More available books at **www.hansebooks.com**

THE LIVING ORGANISM

AN INTRODUCTION
TO THE PROBLEMS OF BIOLOGY

BY

ALFRED EARL, M.A.
LATE SCHOLAR OF CHRIST'S COLLEGE, CAMBRIDGE
OF THE INNER TEMPLE, BARRISTER-AT-LAW

London
MACMILLAN AND CO., Limited
NEW YORK: THE MACMILLAN COMPANY
1898

PREFACE

AMONG the many branches of that organised pursuit of knowledge which has attained in this age its most marked development, there is none which has so profoundly influenced current thought as Biology. The conceptions derived from biological research have proved not only an intellectual gain of remarkable value in their proper sphere; but they have had, so dominant are they in character, far-reaching results in other directions. The great tide of new ideas has been too strong to be contained within its own channel. The neighbouring fields of philosophy have been affected. The human outlook on Nature has been changed.

But just as new ideas colour those which have been previously acquired, so does the loss of a single conception involve the disturbance of many. And, therefore, if Biology is to be a subject of education as well as of research, there are urgent reasons why all who aid in its progress should help to make its foundations stronger, more self-contained, and steadier. Without dwelling on this point, it may at least be observed that inaccurate conclusions and unsound speculation have more serious issues in the case of Biology than in any other of the Natural Sciences.

Security against these results must come about through care in the instruction of rudiments. As far as the intellectual difficulties of the rudiments will permit, a change in the prevalent attitude of the student towards them seems desirable. Too much is taken for granted. The problems and generalisations of the science of living objects are regarded too complacently by the beginner. He is too easily contented with description and narrative; and it may be mentioned that practical work in the laboratory is often little more than an aid in remembering description. That which is a mere statement or a collection of terms in a text-book does not become a course of reasoning because it is followed concretely in the laboratory. The dissection of typical organisms is not necessarily an intellectual exercise.

Moreover, since Biology as the science of living things deals with phenomena of which we all have some kind of knowledge, its ideas lend themselves very readily to popular discussion. It has provided new terms and phrases for a public interested in novelty but impatient of study. Yet the common occurrence of such terms as Heredity and Natural Selection does not imply a widespread extension of scientific research. The respect shown for any generalisation of a chemical or physical nature affords a striking contrast to the confidence with which biological subjects are discussed. It would appear that the problems of inanimate Nature, the qualities and changes of matter, are to be left in the hands of experts who approach them after a rigorous training, while untrained judgment may exercise itself over the mystery of life.

Nevertheless it is with great diffidence that I have ventured on this essay to deal with some of the difficulties of the subject, and to sketch in outline a course of thought which may serve as an introduction to more special study. Detail has been sacrificed in it to breadth of view. Practical illustration is to be found in many excellent text-books, and by the side of those more solid achievements this effort must play a supplementary and subordinate part. The object of the book will be attained if it succeeds, although it may be chiefly by negative criticism, in directing attention to the important truth, that though chemical and physical changes enter largely into the composition of vital activity, there is much in the living organism that is outside the range of these operations.

CONTENTS

CHAPTER I

INTRODUCTORY STATEMENTS

The exercise and end of perception—The nature of knowledge—The nature of observation—Variety of forms in the world—Systematic knowledge needed—Biology deals with living things . . . . . . Pages 1-10

CHAPTER II

SOME CONDITIONS OF ACQUIRING KNOWLEDGE OF LIVING THINGS

Subject and Object—Perception a part of life—The various sciences overlap—Biology deals with feelings as well as appearances 11-18

CHAPTER III

AN OUTLINE OF THE SCHEME OF BIOLOGY

Explanatory remarks—Elementary ideas of Biology belong to two classes—Common forms and changes of life to be described in general terms—Details not to obscure generx THE LIVING ORGANISM

alities—Generalisations simplify description—Changes of matter common to all forms—Description of the scheme of Biology . . . . Pages 19-29

CHAPTER IV

THE STRUCTURES ACCESSORY TO ALIMENTATION IN MAN

General description—The organ concerned in exchange of gaseous material—The organ occupied in distribution and excretion—Mechanical analogies are misleading—The general results of circulation 30-42

CHAPTER V

A DESCRIPTION OF OTHER FORMS WITH REGARD TO ALIMENTATION

Vertebrates—Molluscs—Arthropods—Echinoderms—Simpler animals—Plants—Summary and amplification . 43-55

CHAPTER VI

THE OBJECT OF CLASSIFICATION

Need of an order for living objects—Simple description inadequate—The process of classification—Difficulty in applying the process to living objects—Dual aspect of the organism as possessing form and as presenting change—Classification simplified by regarding both form and function—Idea of development in classification—Conception of types—Influence of authority—Communication of knowledge increases its exactness—Opinions of others to be recognised—Summary 56-79

CHAPTER VII

CERTAIN GENERAL STATEMENTS CONCERNING ORGANISMS AS INDIVIDUALS

Explanatory—Life continuous, Forms intermittent—Form disturbed in reproduction, but life persists—Death the sequel to reproduction—Further discussion of growth—Assimilation the reality underlying growth—Reproduction connected with the process of growth—Permanence of the form—Summary . . . . . . Pages 80-96

CHAPTER VIII

A GENERAL DESCRIPTION OF THE ORGANISM AS RELATED TO ITS SURROUNDINGS

Relation of Subject and Object—Our difference from other organisms in relation to surroundings—One part of life engaged in observing other parts—Natural Science chiefly describes phenomena—Organism both Subject and Object—Organism and surroundings inseparable in some respects—Thought based on sensation, but implying something more—The same world of Objects giving impressions to innumerable individuals—Meaning of "Self"—Perceptions and ideas—Province of Biology—No disparity between organism and environment—A contrast necessary—Link between organism and environment. 97-117

CHAPTER IX

THE MATERIAL BASIS OF LIFE

Introductory—Living changes few in number and forming a constant group—Matter concerned in the changes to be

xii THE LIVING ORGANISM

considered — Difficulty in perceiving the substance as common—Degrees in resemblance of material—Similarity of substance and similarity of behaviour — Process of generalising—Observation needs control and definition—Limitation of chemical methods—Application of chemical nomenclature — Complementary relation of plants and animals — Chemical names given to certain constant phenomena . . . . . Pages 118-136

CHAPTER X

THE ORGANISM AS A CHEMICAL AGGREGATE

Description of chemical changes—The organism as an aggregate of certain kinds of matter—Chemical description not distinctive—Description of matter in general terms—Reality of sensations—Perception of sameness in material objects is exact—All knowledge of Nature based upon the perception of similarity in quantity or quality—Test of sameness applied to living matter—Primary restriction of observation—Protoplasm more than a chemical compound—Constant features of living matter do not preclude variety of form—Difficulty in localising attributes . . 137-158

CHAPTER XI

THE ORGANISM AS A CENTRE FOR THE TRANSFORMATION OF ENERGY

Physical methods and results applicable on certain conditions—Classes of change to which applicable—Conception of energy—Energy of animals derived from plants—The organism not an ordinary conservative system — Resemblance of organic activity to that of other systems—Distinctive features of organic activity 159-174

CONTENTS xiii

CHAPTER XII

CERTAIN ASPECTS OF FORM AND DEVELOPMENT

Need of guidance—Meaning of degree in these matters—Externals of life dealt with in form—Structure and material to be considered—Value of analysis—Structure demonstrates composite nature of activity—Organism together with environment form a system—External changes affecting the organism—Reaction of the organism—Danger of pseudo-explanations—Cell-division incidental to growth and reproduction—Idea of cell to be used with reservation—Products of cell-division not necessarily alike—Prejudice in matters of form and change—Environment assists comprehension of development—Extension of idea of development and environment—Two main paths of development—Hypotheses useful in co-ordinating facts—Ideas of cell-arrangement more useful than ideas of cell—Cell-arrangement less significant than combination of function—Nerve-development connected with highest function—Equilibrium checked by the Organism—Summary
Pages 175-228

CHAPTER XIII

THE MEANING OF SENSATION

Introduction—Illustration of ordinary mutual action—Additional results when one object alive—Effects of external things belong to two classes—Character of sensation—Implication of mechanism in consciousness—The sentient subject 229-249

CHAPTER XIV

SOME OF THE PROBLEMS PRESENTED BY THE ORGANISM

Recapitulatory—Species—Variation and Adaptation—The Cell as Agent—The unity of the Organism . . 250-266

CHAPTER I

INTRODUCTORY STATEMENTS

THE introduction to a fresh study in some cases is a definite event in personal history. It is possibly an occurrence marked by some circumstance of time or place. Certain ideas may be suggested for the first time or, it may be, facts first come within our ken under conditions which render the occurrence memorable. But this is rare.

The acquisition of any portion of knowledge is, in reality, seldom a sudden operation. It is more usually an insensible drifting through a multitude of mental states, and one idea may be the result of many preliminary efforts.

It is difficult to claim for any portion of knowledge that it is absolutely new, and difficult to assign to it an origin or moment of inception. A little reflection makes it clear that most of our thoughts grow gradually out of those which previously existed. The more recent mental conditions, so far as they are progressive, are built up of simpler and earlier states. A gradual progress in thought may often be traced through many stages from an origin which is distant and obscure, though great care and skill are necessary for success in this performance.

This statement, if it can be realised for a single idea, or even for any given class of thoughts, holds still more true of such a vast body of knowledge as is denoted in speaking of Natural Science. There is no absolute and marked beginning of such a subject, as there can be no end, in minds which have ever been active in its acquisition.

We shall learn hereafter that minds must be so occupied as long as they exist. Activity of this kind is a condition of their existence. *We are always learning, though not always are we learning with method or with advantage.*

We perceive Objects and Changes

It is unnecessary to speak of the variety or the number of objects which come within our perception, and the instances of change are innumerable. A complex world of endless *objects* and *changes* appears before us, and our attempts to understand it have created Natural Science.

The Act of Perception is the Beginning of Knowledge

The objects and the changes in those objects, which together constitute our world, give rise to successive acts of *perception*. It may be said that perception is *exercised on objects* (or things), and *on changes* (or events); and each individual perceives within certain limits, and not beyond. Our science or knowledge is based upon, and begins with, the perception of our surroundings.

The Amount of Knowledge depends partly upon the Extent of Perception

This statement must be true, if knowledge be a result of perception. The wider the region surveyed, in other words, the greater the number of objects and changes perceived, and the more accurate the perception, the greater will be the amount of knowledge. The value of knowledge, however, is estimated by very different means. The worth is not likely to be confused with the amount.

The Range of Perception for the Individual is limited

The individual can explore but a limited region of the world. A few alone of its innumerable objects and changes can come within the compass of his means of perception. The experience of other individuals may assist him to widen the circle of his own knowledge. Information gained from others may add to his personal store.

It is the aim of education to put the learner in touch with accumulated knowledge, as well as to render successive additions to knowledge possible. The capacity to perceive may be greatly extended by a careful training which has these ends in view.

The Exercise of Perception needs to be assisted and directed

The mere accumulation of facts, as gained from repeated acts of perception on one's own part, or from statements of perception on the part of others, is not the only end towards which the student of science is directed. Unassisted efforts would probably lead to this result in the main, and then little intellectual progress would be made. Guidance and direction are always needed by the learner, and in almost all stages of civilisation the younger generation has been found to rely upon the older.

The Extent of Knowledge is not so important as its Nature

We do not necessarily know more, because we

have come into touch with a larger number of facts or statements. Knowledge, or science, does not begin until facts are perceived to be *connected* in some way or other.

The facts to be known are without limit. On the other hand, the capacity of the individual to perceive when facts are connected is limited very narrowly. It is important, then, that the working of our perception should take the right direction. The meaning of right, however, remains to be defined.

To observe is to perceive intelligently

It has been stated that ordinary perception, even when the action frequently repeats itself, does not necessarily lead to knowledge. For comprehension of what is perceived, it is essential that the percipient individual should be able to refer the things or events which are before him to that permanent record of past experience which he speaks of as his *mind*.

Perception may be intelligent, or it may be without meaning and without use. When it is directed to a certain end, and when it is consciously performed for definite purposes, it is usual to describe it as an act of *observation*.

Exact Knowledge begins with accurate Observation

There is no doubt that accuracy in personal

observation is essential to progress. Without some training in the special modes of observation which are needed in the branch of knowledge to be undertaken by the student, he cannot expect to be able to interpret rightly the records of previous observers. Their work will not be understood, nor will their descriptions *convey the right meaning* to him.

The methods of acquiring exact knowledge and of gaining a faculty for correctly interpreting recorded facts of observation, based as they are primarily upon accurate personal observation, cannot be given in a few propositions. The following pages describe a few conditions which have a special bearing on Biology.

An Experiment is a special Kind of Observation

When the events to be observed are very complex, it is usual to resort to an *experiment*, whereby acquaintance may be made with a *single aspect* or a *single component* of a given change at a time. All observation partakes more or less of the character of an experiment, in so far as it has ulterior aims.

The Number of different Substances in the World is limited, though the Number of different Forms is indefinite

There is no doubt that the number of different objects even within our own compass is indefinitely large, and within the known world there is no imaginable limit to the variety of distinct things. Yet the many objects coming under our observation are not unlike in all respects. In some of their properties they are frequently alike. Their distinctness or separateness is none the less true because they possess some degree of sameness. The substance or material of two distinct objects may be alike, while the *form, shape, size, situation,* or *state* may differ.

Herein lies the main boundary between the province of Chemistry and that of Biology. Chemistry deals with the composition of bodies, and does not regard their form. It is concerned, too, with the nature of the changes which may occur when different objects are brought together, but it pays no attention to the *form* or to the *identity* of those objects.

We may say, then, that Chemistry is confined to *substance*, while Biology deals with the substance and also with the *form* and *identity* of certain objects.

The term Form may be used to denote Condition or State, as well as Shape and Size

Form is that which is occupied or filled, that which contains, as distinct from what is contained. And so we come to use the word *form* as the sign for the *mode* or *state* in which an object appears to us. The form is opposed to the substance of an object. The matter or substance may be similar in different objects, while great variety of substance may be presented under the same form.

Then, as we find expressed in the science of Chemistry, matter exhibits great variation, independently of difference of form. With what is denoted by the narrower meaning of the term form Chemistry does not often concern itself. The investigation of states of matter, on the other hand, does form part of chemical research, though it is, speaking strictly, the subject of Physics.

The Acquisition of systematic Knowledge is the End in view

The propositions which have now been laid down form a suitable introduction to our subject. They have no special reference to any particular branch of natural science, but they apply to all knowledge derived directly or indirectly from *observation*.

In connection with Biology, however, they are apt to be forgotten. In that science it is more than usually important to start in the right direction, and, indeed, to maintain throughout a correct attitude with regard to the foundations of our knowledge of Nature. For this reason, after recapitulating the gist of what has been already stated, we shall proceed in the following chapter to extend our list of preliminary statements.

We derive our knowledge of Nature from the exercise of observation, an act which has two component parts—*the normal activity of the senses* on the one hand, and *the direct control of the intelligence* on the other. The mere perception of the objects and changes among which we live is not enough for reasoning beings. It is not enough to receive impressions of nature as a series of separate and distinct phenomena, nor to receive them as a crowd of confused images.

The acquisition of systematic knowledge exacts much more than a passive submission to external influences. It is a process evidently controlled and directed to a large extent by the experience accumulated in the past and made accessible through records. One's own observation does not bring to light the *connections* between phenomena and the *orders* among events, until it is aided by the experience of others and shaped by reflection.

Indeed any course of study is only to be distinguished from the daily round of ordinary watchfulness and reflection by the fact that these activities, which are evoked in some degree from every individual by his surroundings, are therein *directed* and *controlled* to a definite purpose.

And with such a purpose do we undertake the study of Biology.

Biology is the Science which treats of living Things

Such a statement as this will scarcely satisfy the inquirer, for it does not tell him enough. As to what is meant by *a living thing*, there is indeed a general, though indefinite, consent, and it is sufficient for the ordinary pursuits and thoughts of most people. *Animals* or *plants* as examples of living things are easily recognised. Mistakes are made, it is true; but, as a rule, the thoughts we have about animals and plants are precise enough for ordinary purposes. To be more precise or more definite, our ideas would need an amplification which must be the result of guidance and training.

Although it may be useful at this stage to make a more exact statement of the nature of our subject, it will not be till the end of our course, in default of some previous training, that the meaning of the statement can be comprehended more definitely. It suffices now to say, that *the science of Biology is a certain body of connected knowledge, which is derived from the observation of living objects and from reflections based on that observation.*

CHAPTER II

SOME CONDITIONS OF ACQUIRING KNOWLEDGE OF
LIVING THINGS

The special Difficulties of the Subject are great

THE material of study may be named, and even the end, but we must be content to give to the mental processes which are concerned in the acquisition of all exact knowledge, whether biological or not, nothing but the shortest of descriptions.

Mental changes are difficult to follow. We cannot, for example, understand what is meant by a *generalisation* or an *abstract thought*, till our minds have learnt how to generalise, and we have been conscious of many an abstract thought. To attempt to give a definition of a thought so abstract as that represented by the word *life* would be useless. Too wide a range of experience would be needed to grasp the full meaning of such a word.

The forms which exhibit life, and the activities

which represent it, must be first observed and pondered over, both separately and collectively, before we can gain an adequate conception of life as a whole. That our ideas about life are limited will be apparent at an early period, and there will be no stage at which our mental activities are not confined at some points by barriers which cannot be passed.

But it is not so much the boundaries set to our thoughts that prove the first impediment to progress. It is rather the vagueness of the thoughts themselves. The more familiar examples of living things may be, it is true, the subject of certain propositions more or less valuable. But how is it possible that *all* the varied forms and events which come before us can be covered by the single word *life* ?

It is not merely the innumerable *individuals* nor the almost innumerable *kinds of individuals* which bring about our perplexity. We are troubled by the difficulty of incorporating the multitudinous forms and the varied changes in some comprehensive statement, the difficulty, in fact, of finding anything which applies to all cases and may be expressed of all alike. There is in some instances indeed little to be perceived, save an indistinct manifestation of an unusual movement or an organic shape.

When any part of living nature confronts us, it needs no little skill in unravelling our thoughts and no little patience in observation, to gain consistent ideas; and, still more evidently, the best powers of the intellect will be needed before any *order* or *arrangement* can be given to the total sum of living things. The attainment of an order out of the apparent confusion has been the result of many generations of thought. Among the products of that achievement are the following elementary ideas.

Perception implies the Existence of a Perceiver

These words convey a truism, which hardly seems to need expression. Yet it is given in order to emphasise the fact that perception is to be regarded as an event which has two components. The thing perceived and the person who perceives stand in apposition to one another. There is *a physical universe* on one side, and *a being sensitive to it* on the other side.

The Thing perceived is the Object. That which perceives is the Subject

This is but a repetition in other words of the previous statement. A name is now given to each aspect of the act of perception. The subject *perceives;* the object *is perceived.* It is advisable to be content with this statement for the present, without attempting to grasp all that it may suggest. Further consideration would be interesting, but it would lead us away from Biology.

Perception implies Life

The ability to perceive or to receive impressions is clearly a character which belongs to some living objects. It cannot be said with certainty that the character is one which always distinguishes living from inanimate things, until we have decided what things are alive and have learnt more about perception itself from our own practical experience.

It should be noticed, however, that the term *subject* has, so far, only been used with reference to an animal, and a rational animal. Whether all forms of life can exhibit the same relation to external nature as is implied in the use of the complementary terms, subject and object, remains a question for later consideration.

The various Sciences encroach on one another, and in Biological Investigations the Elements of several Sciences are involved

The full discussion of the matter on which we have entered involves the study of other sciences. Psychology, for example, is one which specially deals with *mental processes*, resolving them into their elements and tracing out definite stages in the operation of thinking. It is a science which is occupied wholly with those facts of life which are described as *subjective*, namely, *Consciousness* and *Thought ;* and even a slight acquaintance with such facts, with their origin and their analysis, will prevent many disastrous blunders in our biological conceptions. It is hoped that certain ideas, which belong strictly to Psychology rather than Biology, will be conveyed in these pages to the advantage of the student of the latter science.

The Sources of Error are much the same in all Branches of Science

But difficulties and errors are not confined to any one branch of study. They are common to all, and in all they are chiefly due to the same causes.

Preconceptions and erroneous judgments are foremost in retarding progress, and with these, as a rule, beginners in Biology are plentifully endowed. The thoughts which are aroused in their minds by a given term do not always agree with the thoughts which are truly represented by that term. From such faults incorrect interpretations of description arise; the record of a fact fails to recall the fact where such defects exist.

By a constant reference to definitions, probable errors of judgment and imperfect generalisations may be avoided in many cases. But the form in which this subject is introduced, and the length of our preliminary discussions, are dictated by consideration of the exceptional difficulty in avoiding *hasty generalisations and unwarranted assumptions* in dealing with life.

The Study of Life should include Feelings as well as Appearances

Finally, we may assert that the acquisition of knowledge varies very little in its methods. The mental faculties which are employed in Biology are called into use in other branches of science. It is evident from our previous statements that some attention must be given to *thoughts* and *ideas*, as well as to *objects* and *changes*. That which we ourselves think about the world, and perceive in it, must be part of our study. And it is best to realise this at the beginning, before trying to form judgments even about the simplest appearances of life. It would lead, more certainly in the case of Biology than in most branches of science, to incomplete, if not to erroneous conclusions, to ignore *the relation of the observer to the thing observed*. Yet the full meaning of that relation can be but imperfectly understood, even at the end of our course.

There are, then, two classes of things which must be admitted to exist beyond doubt, namely, *Feelings* and *Objects*. No exercise of thought ought to lead to a confusion of these two classes. They are esteemed by universal consent to be separate and distinct in our consciousness. We accept as the primary facts of our life the existence of a manifold world of feelings and thoughts, forming what we call Mind, together with that of a complex world of objects beyond and external to that mind.

Feelings have always to be referred, directly or indirectly, to a world of objects, and yet they must be regarded as having an existence apart from those objects. Consciousness takes this form most readily. Whether the judgment implied by the distinction is a *final one* depends upon the individual. For most people it is final.

But it may be asked, What has Biology to do with this fundamental generalisation, that there is a *Self*, or perceiving mind, and a *Not-Self*, which is the world of objects distinct from the perceiving self? It will be the aim of the following pages to show that reflections which arise from a consideration of the relation of Self and Not-Self, or, as it may be termed more fittingly, the Subject and Object, are essential to an adequate grasp of biological questions.

Further, we may say that the world, which is apparent to us as external, everywhere presents itself as a diversified substance in incessant change. At the same time, it is evident that without some permanence of substance, or constancy of form, we should be unable to detect change of any kind.

Our conception of life is necessarily associated with the image of a certain permanent form of matter undergoing certain modes of change. Our conception of life is limited, in other words,

C

no less and no more than the rest of our very conditional and relative knowledge.

Lastly, it may be repeated, we can observe life in others and also feel it ourselves; but though the life of others appears to us as definite changes taking place in diverse forms, it is meaningless to speak of feeling or observing the feelings of other living objects.

CHAPTER III

AN OUTLINE OF THE SCHEME OF BIOLOGY

Explanatory Remarks

THE statements to be subsequently made about the organism will be found simpler in kind than most of those already encountered. They describe the more prominent appearances, or what are commonly called the more striking facts of life. These appearances are presented to all of us in the same fashion. The mode of presentation—*why this or that thought should be aroused by a given object*—is a question not touched upon in this chapter.

Yet the resemblance of ourselves, the observers, to the objects we are observing must not be forgotten. So much in common exists, that there is a danger of considering everything to be alike. It is advisable on many grounds to give attention to this tendency, and there is no doubt that it is an influence in many of our judgments concerning the nature of life. *Ideas derived from our own experience of life are too readily transferred to other forms of life.* The resemblances are apt, indeed, to disguise the differences.

As already stated, the mode of treatment employed in these pages has taken its shape from these considerations. It has mainly originated in the effort to separate *assumptions* from *truths*, and to show that the thought which is a part of our own life is occupied in forming conceptions about life as a whole. Hence it is that mankind is only in appearance occupying the first place in this essay.

It is necessary to deal with some of the conditions of human consciousness at the beginning, lest our incomplete conceptions should rank as truths and many difficulties be overlooked.

The elementary Ideas of Biology belong to two Classes

Our earliest observations and our earliest information in this science will give rise to impressions which fall naturally into two groups. There are, standing in marked contrast, two main categories in which the ideas of Biology may be placed. It will seldom happen that there is any apprehension of physical life which does not come under the heading either of " *Thoughts about Form,*" or " *Thoughts about Changes.*" In other words, the material of Biology suggests *a clear separation of the facts of form from those of change.*

Yet it must not be supposed that the two

classes are unrelated. Hereafter it may be important to treat the two classes of facts together, while at the present stage it is a great convenience to distinguish them in our minds. It is an advantage to place on one side the impressions derived from a multitude of living and material objects of very varied shape and structure, and to keep them as far as possible in the background, when ideas of the changes associated with life are occupying our attention. During the progress of observation and in the course of description, forms and changes necessarily come together. *It is in the process of reflection and during consideration that they need to be separated.*

The common Forms and Changes of Life must be first described in general Terms

The *forms* of life are beyond enumeration, and the illustrations given below are a few in a multitude. The *changes*, however, which living forms exhibit are quite limited in kind. The most *prominent* are changes of position, the most *universal* are those connected with feeding, and these two classes of change will be the first to be described in outline. It is in connection with such events that some description of the forms of life is most fittingly introduced.

The ANIMAL KINGDOM, as it is called, is made up of individuals which are in most cases capable of locomotion. The VEGETABLE KINGDOM, so called, is distinguished by immobility.[1] All members of both divisions are alike in their dependence upon certain external matter, which we call food. The act of feeding is a necessity for every living object.

But it is a very short step towards an arrangement of these living objects to divide them into *animals* and *plants*. Sub-division after sub-division has yet to be made. The dwellers in the sea or air differ from those which live on land, and the differences are connected with the habits which are imposed on them by the nature of their surroundings. We may, or may not, perceive *organs, limbs*, and *symmetry of outline;* for, indeed, a definite shape cannot always be predicated of the organism.

On the other hand, we may note in many cases a rigid symmetry, and for those of the same class an undeviating sameness of outline. Yet these points, if we be inclined to deal with them by themselves, are not significant enough, nor of sufficient interest, to occupy our whole attention.

Details must not be allowed to obscure Generalities

We are almost compelled, by force of numbers and heterogeneity, to examine a few specimens of living forms rather than attempt a review of a larger number. Yet our comprehension of the object of study *depends as much upon our capacity*

[1] A useful, though not a strictly accurate distinction.

to generalise as upon our skill in dissection or analysis. The knowledge of details of structure, and of the countless variations of strncture, does not become coherent or edifying, unless it be acquired as part of a wider scheme of perception, —unless, indeed, it be conceived as subordinate to larger principles. With the object of learning to generalise, we shall proceed to the description of certain facts of common observation.

All *beasts, birds, reptiles, fish,* and the innumerable *invertebrates,* together with every kind of *plant* or *tree,* yield, each in their own way, some quality or property apparent to our sense of sight, which permits not only a ready distinction of species from species, but also the grouping together of numerous individuals in the *same class.* But where are we to begin? Of the many appearances presented to us in the web of life, which will best serve for purposes of classification, or which will give us the best or the widest information?

To enumerate all the appearances or properties of an inanimate object takes time. To give an approximately complete account of all that is to be observed in a living object is a task of much greater difficulty. Even the first attempt to carry this out will show that a bare enumeration of properties is, in itself, not enough. The various modes in which *properties are connected,* and *the association of groups of properties,* will be found to be more significant and more fruitful in information. If we do not limit our activity of observation by considerations of this kind, and do not learn to combine reflection with perception, we shall certainly not attain to much knowledge of our subject.

Generalisations relate to Aggregates and simplify Description

In the ordinary exercise of perception very similar results are frequently obtained. Impressions made upon the percipient subject are often alike in all circumstances but that of time; and the external objects, to which these similar impressions are referred, may be, in that case, mentally grouped together. When occasion arises, they may be *denoted simultaneously by a single name*. Such a name or sign of a number of objects is spoken of as a *common name*.

The mental activities which result in a common name or a common idea being called into existence constitute the process of *generalisation*. To generalise, therefore, is *to create. or receive an idea*, either simple or complex, *which may be applied to a number of objects with equal fitness*, when once it has come into existence. The fitness of the application will depend upon the accuracy with which the general idea has been formed, whether or not it has been based upon adequate observation.

The aggregate of objects forms a *class*, and since it is thought of as a whole, by reason of all members of the aggregate possessing some property or quality in common, we are able to *abridge our descriptions* in using the name of the class. We no longer need to specify this and every individual in recording our observations, but our common name or general statement embraces an indefinite number of instances in a general proposition or generalisation. Description is shortened by this means wherever resemblances occur, and the range of thought is extended by economy in expression.

In the following outline of some of the prominent facts in the ordinary aspect of life, the use of generalisation is apparent. The condensation of description is evident, and the acceleration in mental processes which follows the conception of classes is illustrated.

Constant Changes of Matter are common to all Forms

No living object, though it preserve its identity in some respects, is able to maintain the same material composition for long. *Its form, though relatively constant, has a continuous stream of material passing through it.* This fact is often expressed by saying that there are two operations, *waste* and *repair*, in continuous progress. But it is doubtful whether anything is gained by the use of these metaphors, and there is always a risk of misconception through forgetting that they are metaphors.[1]

The most careful and extended observation has failed to detect any exception to this general statement, and it remains as a distinctive character of life, *that material changes are essential to its persistence.* The process of generalisation cannot be pushed further in Biology, when once it has embraced every object included in the range of the science. But it is well to remember that the limit of certainty is never actually reached. And at the most we can only say of the proposition before us that there is nothing in the collective experience of mankind to contradict it.

The Generalisation is so universal that it ceases to be instructive

The *limit* of generalisation, however, is not the sole aim of scientific inquiry. Ideas of narrower range have first to be acquired and formulated. It is really more important in the present instance to be able to discern differences, rather than resemblances, in the manner in which this change of material is carried on.

In fact, our knowledge of living characters is more likely to be increased by comparing different

---

[1] A metaphor is a word used in a sense which is not usual, *i.e.* with a meaning not proper to it.

instances of each distinct mode of change than by struggling after the limit of generalisation.

Description of the Scheme of Biology

The vague description of Biology previously given, as the science of living objects, will no longer be found exact enough for our purposes. There are a number of sciences which occupy themselves with living objects, *Botany, Zoology, Physiology, Morphology, Embryology*, as well as others. The chief object of Biology is to serve as an introduction to any of these branches of study. But it will not serve its proper purpose by expounding the rudiments of these subjects in succession. Something more than this is needed.

From the facts and generalisations accumulated in the various sciences of life, certain principles have been abstracted, and with the exposition and the application of the principles so derived Biology is almost wholly occupied. These principles are drawn from two, or at most three, prominent conceptions, to one or other of which all the sciences of life contribute every acquisition they make. They are the conceptions of *form, function*, and *development*.[1]

Every fact of life comes sooner or later under one or other of these headings.

[1] Development may be omitted as being a derivative of the other two, see chap. xii.

It is, therefore, essential that the facts of life which are to enter into the composition of Biology must do so as portions of a scheme. Facts in isolation may legitimately enter into the records of other sciences, but it is only *in their relation to one another* that facts are useful to the biologist; for he is attempting to learn, above all things, the fixed principles which underlie the diversity of life. He is endeavouring to discover the elements which combine to form, and through which he can understand, the complexities of Physiology, Zoology, and Botany.

To gain this end, an abstraction or general conception of the organism has to be built up, however difficult the process may be. So, too, *form* and *function*, *structure* and *change*, have to be treated in a broad manner, while *the individual case* is but introduced *in illustration*, and not for its own sake.

As far as possible, the attention should be directed to those phenomena in the organism which are prominent and universal—that is, manifestly exhibited by all. In fact, *those qualities which are not modified by form* claim our first care. But to succeed in finding out what is always present, we have to learn how to pass by much that is superficial and distracting, and how to distinguish that which is really essential from that which is merely familiar. Many gradations likewise exist in the importance of qualities, and many steps remain to be taken before these can be discerned.

All the discipline here planned is preliminary to the stage of research into ultimate causes. *Why the form is thus, the function or the development of this or that kind, are problems of much later date, and Biology cannot be expected to answer them with certainty.*

CHAPTER IV

THE STRUCTURES ACCESSORY TO ALIMENTATION IN MAN

General Description

A SPECIAL structure, the alimentary canal, serves in man for the reception and conversion of food. The food is progressively altered during its passage through this tube, by coming into contact with special portions of the body which react with it. More or less changed by or combined with these substances, the assimilable portion of the food afterwards enters into the material constitution of that body. But the incoming material does not *permanently* reside within the boundaries which we recognise as the form of the individual.

A concurrent loss of matter is always going on. Even in the case of an individual rapidly growing, there is evidence that the total quantity of food taken during any given period exceeds the increment in mass which is gained. The longer the period over which observation extends, the more convincing is the evidence. In the case of the mature individual, the loss which is indicated by this fact is closely balanced by a gain, and, though undergoing minor changes periodically, he maintains a fairly constant *average* of mass.

As far as can be observed, the contrary streams of loss and gain of matter are universally associated with living changes. They are encountered in every form of life, whether *animal* or *plant, large* or *small, simple* or *elaborate.*

But it will be observed that there exists in the structure which is organic to this end, a distinction of parts ; and it is a distinction which varies according to the kind of organism under observation. In man, at the entrance to the canal is an organ, the *mouth,* provided with sharp and hard teeth (fixed in a framework of bone which admits of a certain degree of movement). By aid of this apparatus an action of cutting and tearing takes place previous to the disintegration, which is afterwards brought about by the presence of *solvent liquids.* Some liquid is introduced together with the solid food, but the liquids which are most active enter the canal from the various glands engaged in secreting them.

Certain of the glandular structures, such as the several *salivary glands,* the *liver* and *pancreas*, are comparatively large organs contributing a great quantity of active material towards the process of digestion. But liquids of various kinds and functions pass at different points into the canal, and throughout its whole length there are present small structures, evidently capable of assisting in finally bringing about a close contact of the nutriment with the living substance of the body.

Now the whole of the glandular apparatus is *resorptive*, either directly or indirectly, as well as *secretory*. The stream towards the canal is at least equalled by that flowing away from it. To convey a more exact picture of the changes in alimentation, we should need to describe them as a system of *circulation*, into the various channels of which new material is being drawn, while other material is being expelled from them.

Every portion of the glandular system, with its accessory network of blood-vessels and lymph-channels, takes its meaning from this function alone, or, as may be said, it is *organic* to this single end. Even the more distant glands may be regarded both structurally and functionally as part of the alimentary apparatus.

External to those absorbent and active layers of the canal which are more directly concerned in the exchange of material, we find a layer of muscular bands so disposed that a contraction of successive portions of its length may take place. The necessary contact of the nutrient material with the lining of the canal, and with successive portions of the same, is thereby made more complete, and the expulsion of material which has not been absorbed is facilitated.

The Organ concerned in the Exchange of gaseous Material

Adjacent to the alimentary canal proper and closely connected with it are the *lungs*, which are truly alimentary in function, though commonly spoken of as the *respiratory organs*. In position, growth, and function they are allied with the alimentary canal. There is, in fact, a common purpose served by both—*the interchange of external matter with that forming part of the organisation of the body*. The lungs, however, are wholly occupied in the exchange of *gaseous* material; while solid and liquid exchanges take place through the canal alone.

The structure of the lungs permits a ready absorption of oxygen from the inspired air, by the agency of a fine network of capillary blood-vessels, which are distributed throughout its sponge-like substance. The air enters by the *trachea* through the *bronchi*, and thence into the innumerable *air-cells* which give the lungs their peculiar texture. A thin partition alone separates the air at any point from the blood, which circulates through every portion of a crowded network of vessels, and a rapid and voluminous exchange of gases is thereby rendered possible. As a final result, the blood becomes richer in oxygen, and the expired air shows an increased quantity of carbon dioxide.

The Organ occupied in Distribution and Excretion

The *circulatory* or *vascular* system forms the apparatus by means of which material from outside is distributed to every part of the body, and portions of the body are rejected or excreted. Although a considerable amount of matter leaves the body by way of the alimentary canal, it must not be forgotten that every particle of matter which is truly excreted, as distinguished from that which is merely unabsorbed, has passed through some portion or other of the *vascular system*.

The process of circulation is maintained by a central propelling organ, the *heart*: *distributing vessels*, either arteries or veins, and various irregular cavities and channels; while certain liquids, *blood* and *lymph*, with their complex ingredients, constitute the ever-changing material in circulation.

The heart, which is a localised muscular enlargement of the vascular system, acts as a propelling and an exhausting organ, driving blood through the arterial system by muscular contraction, and assisting its return through the veins by the relaxation of its muscles. The construction of the chambers of the heart and the arrangement of its valves promote this work.

The flow and return of the blood, to and from all parts of the body, is necessary for the work of uniformly distributing the matter absorbed from the alimentary canal and the lungs. The branching of the larger vessels into smaller and smaller channels finally ends in a minute capillary network, which is to be met almost everywhere in the body. But the minute vessels and spaces, which constitute the network here described, begin after a time to reunite into larger and larger vessels. An inverse process is encountered. The channels soon become larger and fewer, and finally the blood re-enters the heart by a single large vessel, and begins its course anew. The construction of the system, in brief, enables the blood to be rapidly driven through every part of the body, and, as rapidly, collected together again.

This bare description, however, merely gives us information, and that of a superficial kind, of the changes *preliminary to assimilation*. It gives something of the plan in outline, omitting altogether the subordinate movements of the blood, and of its accessory liquids, without which the general circulation would be ineffective. But the plan, even if complete, would include few of the serious difficulties of assimilation itself. These structures are but part of the coarser mechanism which is engaged in the antecedents and consequences of a more subtle act.

We may state that the function of the vascular system, as a whole, lies in the *preparation, distribution,* and subsequent dispersal of all the matter associated with the organism. The substances to be found in that vascular system, in some part or other, typify all the changes of form and composition which are connected with the continuance of life. To regard the system as simply occupied in the distribution of ingested food would be to form a judgment distinctly erroneous. Distribution is part only of its function, for that which is being distributed is meanwhile being significantly transformed.

The substances contained in the vessels which leave the heart undergo a *succession of varied changes.* At frequent stages in its course, the blood receives additional material, and also suffers loss. The sequence in states is as important as the stream of material.

The structures which contribute to these changes in composition are, in the main, the *lungs, liver, kidneys,* and the various *glands* which have been already mentioned in connection with the alimentary canal.

The vascular system, then, does not simply carry and allot the matter taken from the alimentary canal. It receives it by way of the *lacteals,* brings it under the action of living products at once, and continues to submit the material to a series of operations before finally localising it. A broad survey would fix the food, which has been absorbed after a series of preliminary changes from the alimentary canal, *at one end* of a lengthy process, and the arterial blood, as it is driven from the left ventricle, *at the other end*. These are the limits of a series of chemical and mechanical events which form an important part of vital changes. *Between these two limits that which is most remarkable among many occurrences is the preparation of inanimate matter for the inception of life.*

Mechanical Analogies are misleading

There is much to be gained from the judicious use of *analogy* in attempting to describe complicated occurrences. An analogy is often helpful as an ideal framework, on which detached thoughts may be brought into relation with one another.

In illustration of this auxiliary of information we may compare the food of an organism with the fuel of an engine. The material necessary to the life of an organism resembles in some respects the substances which originate the movement of machinery. But just as the life of an organism is far more complicated than any activity which can be manifested by machinery, so the events which take place in alimentation are far more complex than those chemical and physical changes which supply the motive power of an engine.

The resemblance between the two lies mainly in the dependence of each for its activity upon a continuous supply of suitable material. Here the analogy holds. But in other respects the differences are so great that the analogy is likely to be misleading. Its application would be extended if we knew of a machine which is not only able to *supply its own fuel* and adjust its activity to varying circumstances, but also to incorporate with its own substance the fuel which supplies this activity. The food of an organism may be looked on as its fuel, but it becomes also the machinery which is at work. *That which is consumed in the case of an organism is the machinery*, and food repairs the gaps left in a mechanism.

To be accurate in the use of our analogy, we should compare the visible substance of the organism with the fuel rather than with the machine, for that is the substance which is being consumed. But in no case has the food itself any real parallel in the fuel of an engine. The material must be radically changed in form before it can serve as fuel.

Where now do we find the machine? What is there in the organism comparable with the combination of resistant bodies, which we recognise as a machine when we perceive its parts to be capable of certain relative motions? *Design* or *adaptation* is evidently common to both. But the working material of the one is characterised by constancy. *The substance of the machine is always the same, the substance of the organism is in constant change.*

The general Results of Circulation

But though it may be impossible to find among the complications of art or nature an occurrence resembling alimentation in many of its aspects, we may render description more easy, if not more intelligible, by noting the most marked stages in the process. There is first *a selective and absorbent surface*—that of the canal, and a supply of food in its neighbourhood; and then *selective and absorbent tissues*, forming the bulk of the individual, and having a supply of previously prepared substances always in proximity. The absorbent tissue, together with its prepared nutriment, is to be found in every part of the body. The surface at which external matter is absorbed is limited. The medium which permits the necessary transfer of matter from place to place is the *blood;* and it is in the blood that the observable alterations in the condition of the matter take place. To bring about these changes there are temporary resting-places in the form of *glands*, where *losses, gains,* and *rearrangements* of material take place.

The glands may be reasonably described as resting-places, because they all exhibit a structure wherein the circulating liquids may be submitted, for a comparatively long period, to the action of the living material of the glands. Blood is supplied to each gland by a network of vessels, ramifying in its substance; and material is secreted by the gland, in quantity corresponding with the supply of blood and in nature dependent on the *specific activity* of the gland. A retardation of the circulation here takes place, from the narrowing of the individual vessels as they branch out and increase in number; and a loss ensues from the process of *selective diffusion* through the substance of the gland itself.

The matter resulting from this species of filtration is either added eventually to the general stock of matter in circulation, or it passes away from the body, as occurs by way of the *lungs* or *kidneys*. The material may be rejected—that is, *excreted*; or, after passing through intermediate stages, it may again join the stream in circulation to undergo further change (sometimes even allying itself with food at its first contact with the body,[1] and afterwards retracing every step that is known to exist in the course of alimentation). In such cases it is described as *secretion*. *The main course of circulation has, therefore, minor systems of circulation in subordinate connection with it.*

[1] As occurs with saliva.

The most general description of the changes briefly recorded above, and of the localities in which they occur, is that which will give most assistance in the attempt to understand their connection with life. Our efforts will always succeed best when they take the direction of expressing observable relations in the *simplest* terms to be found. Only then do we extend our comprehension. The main condition of success, it may be repeated, is to make every statement as general as possible. The description which has been given of the changes occurring in glands has taken its modified form from this motive.

These changes in the form of ingested matter proceed simultaneously with life. They are encountered nowhere but in living organisms, and the whole of the substance manifesting life is undergoing change of a special character, which is too complex for analysis at this stage. *Change is in progress throughout the organism; assimilation and dispersion are the material constants of life.*

If we attempt to fix *limits* to the act of life and say that it begins at this stage of change or the other, we are involved in difficulty. In the end, all we can say with certainty is that each organism is a centre at which inanimate matter is made to exhibit the attribute of life. *The inception of the changes would appear, therefore, to be due to something external to the substances in which they are perceived, while the initiation of life itself is in reality nowhere apparent.* There is no *moment of time* or *position in space* at which any of these changes may be detected *except life be already there;* nor, on the other hand, is there, in the series of events which every particle of living matter passes through, any one to be selected as a sure sign of *its entry upon life.*

CHAPTER V

A DESCRIPTION OF OTHER FORMS WITH REGARD
TO ALIMENTATION

THE illustrations which have been so far given are drawn from observations in human physiology. It remains to be seen how far the principles elucidated may be applied to other organisms, and how far the structure of the latter is conformable to the generalisations about life drawn from a single class.

The characters described are not found to be confined to man. In those animals which readily admit of examination the plainest indications are given, that ingested material undergoes changes and, in part, incorporation with the organism, as it does in man. The need of food and the presence of structures for its reception and transformation are, indeed, almost always evident.

Vertebrates

A similarity in the plan of the digestive apparatus is in itself enough to give a family likeness to the *mammalia*, a large class of animals widely divergent in many other respects; while a still wider group of animals, which are alike in the possession of a vertebral column or backbone— *mammals, birds, reptiles, amphibians,* and *fishes,* for example—have also in common most of the organs which are noticeable in man as accessory to assimilation.

All vertebrata have a distinct and specialised blood-vascular system. A chambered heart is always to be found maintaining a respiratory circulation, though it is only in some of the vertebrates that it influences the larger or *systemic* circulation. A *portal* circulation without exception exists, the venous blood passing through a large gland, the *liver*, which brings about important modifications in that liquid. Part of the contents of the alimentary canal pass through a "lacteal" system of vessels on their way to the blood-vascular system proper. *Renal* organs of secretion are invariably connected with the circulatory system, and either *lungs* or *gills* serve for the exchange of gaseous matter, of which oxygen is the important item.

It is scarcely a matter of surprise that the resemblances with regard to one feature of organisation should find their counterpart in others. The striking *similarity of plan* to be observed in the *skeleton, sense-organs, limbs,* and in the disposition of the *nervous system*, though not entering into the course of this argument, cannot fail to confirm the opinion that the main changes in various species have much in common, if direct experiment had not already produced that result. Yet special differences in structure, and minor differences in the corresponding functions, may be noticed, or the variation expressed in *classes* and *orders* would not be formulated.

Among the most prominent distinctions it may be stated that *fishes* and *amphibians* are characterised by the possession, during the whole or part of their lives, of *gills* or *branchiæ*, organs which permit of the absorption of air dissolved in water.

Birds and *mammals* are distinguished from the latter by having four-chambered hearts to provide for the control of separate pulmonary and systemic circulations.

Teleostean fishes exhibit a number of tubular appendages at the commencement of the duodenum, called *pyloric cæca*, and these share in the process of digestion. Other classes of fishes are provided with an extension of the absorbent alimentary surface in the form of a *spiral valve* in the large intestine.

The respiration of *birds* is remarkable for its extent. In consequence of the admission of air, not only into the lungs but also into large spaces in various parts of the body, a proportionate extension of the surface of aeration ensues, and the more rapid chemical change taking place in the blood is marked by a higher average temperature.

Molluscs

It is not necessary to state that vertebrate animals are very far removed in structure from other living objects. The differences existing between a mammal and a fish are negligible in comparing either with an invertebrate animal. The Mollusca, for example, possess a long alimentary canal, but in them it is simpler in structure, and not enclosed in the general body-cavity. Yet salivary glands and liver are present, in their case also, to assist in the process of digestion. The vascular system is less clearly defined than in vertebrates, though a definite systemic heart is present. Definite organs of respiration may be absent, or a portion of the mantle may be adapted for aeration. Ctenidia are also of frequent occurrence. In the *pulmonata* the mantle is folded so as to provide an air chamber, into which air may pass by an aperture capable of being periodically opened and closed. Excretion takes place by *kidneys* or *nephridia*, as well as in respiration and by the alimentary canal.

The general structure of the body is wholly distinct from that of the vertebrates; as a rule this is most clear in the absence of a skeleton and of appendages for locomotion, and likewise in the development of the nervous system. Yet resemblances are manifest, most prominently in the definite apparatus of digestion and in the vascular apparatus for distributing the digested material, and, less prominently, in the material of the supporting and muscular tissues.

Arthropods

All the segmented animals or Arthropoda—*insects, crustaceans, centipedes, scorpions, spiders,* and their allies, are alike in other respects than that of segmentation. They all possess well-developed alimentary canals with similar accessory organs. Their circulatory system is also constructed on the same plan. It is provided with a contractile heart and a fairly definite vascular system, which are placed dorsally. Respiration may be carried on through special organs called lung-books, or by involutions of the integument called *tracheæ*, but it is also effected in some forms by more specialised organs, the *gills*.

Echinoderms

Among the Echinodermata we first come in contact with forms which do not possess an alimentary canal *traversing the body*. All possess a canal, but in some forms there is but one opening. The body-cavity is filled with a watery fluid, in which float bodies resembling the white blood-corpuscles in animals of more complex organisation. This fluid may be regarded as the equivalent of blood.

A prominent feature in their structure is the *water-vascular system*, the circulation in which is probably attended by respiratory and locomotory results.

The development of both circulatory and alimentary systems is less marked among these forms, and is just sufficient to promote a general diffusion of material through the substance of the body.

The simpler Animals

In several classes of *worms* there are definite digestive, vascular, and excretory structures, though in the flat-worms (*platyhelminthes*) there is no real vascular system, and the alimentary canal has only one opening.

Animals still less organised are included in the classes—*cœlenterata, porifera,* and *protozoa*. Cœlenterata and Porifera possess a definite shape and a digestive cavity. The whole lining of this cavity ingests food without any differentiation of parts for that purpose. In some forms there are *canals*, affording to the organism a larger surface of contact with the water, in which it lives and from which it derives nourishment.

Protozoa, on the other hand, do not always have even a *constant form*. They appear as very microscopic masses exhibiting irregular movement. The semi-fluid mass manifests some structure in the appearance known as the *nucleus;* and the *contractile vacuole* may render service in excretion. There is no cavity, but ingestion may take place at the surface, either irregularly or over a fixed area. The exchange of material appears simple and direct, there is no difference of parts to denote a division of labour, and yet this is still consistent with the existence of life. *The apparent simplicity in form and structure does not, however, render the process of alimentation necessarily more simple, nor the problem of life more easy to solve.*

Plants

We pass now to the other half of the organic world, *the vegetable kingdom*, the life of which appears to differ so much in character and in extent from that of animals, and also in its chemical and physical accompaniments.

The food of animals is always in great part organic. Plants live on inorganic material, *nitrates*, compounds of *ammonia, carbon dioxide*, and similar substances, which do not support the life of animals. Hence it arises that animals are ultimately dependent for their existence upon the material continuously provided by plants.

With few exceptions the food of plants is in the fluid state. It exists as gas in the air, or as dissolved matter in the soil. The leafy expansion of the aerial surface, and the extension of the sub-terrestrial portion of a plant by means of roots, facilitate the absorption of the diluted nutriment in sufficient quantity. In most cases, a system of vessels permits the passage of the liquids absorbed by the roots, and assists in its distribution. The porous structure of the leaves admits the important food, *carbon dioxide*, to their interior, and, at the same time, allows a gaseous excretion or transpiration to take place, the excreted material being chiefly water vapour and oxygen. The process of excretion is limited to this expiration of oxygen and water, together with a small proportion of carbon dioxide, and the continuance of life is therefore marked by an accumulation of material. The tree grows in bulk year by year.

In structure and in appearance the animal and vegetable portions of the organic world have little in common to suggest that they are alike in the attribute of life. But though their food is so dissimilar, they are alike in their dependence upon their environment for a continuous supply of food; and, when once assimilated, the living material of organisms of either class presents little variety in chemical composition. It is therefore reasonably surmised that there is *something in common* between the two kinds of life, different as they may be in most features.

A Summary and an Amplification

The most patent characteristic of an animal is its movement, yet not the fact of movement alone, but the nature. The most obvious test of the life of an animal is its capability of *spontaneous*[1] movement. All animals which come under ordinary observation display at times irregular and fitful changes of position. These changes may or may not be connected with other changes external to the animal.[2]

But there are many other objects which are spoken of as living without exhibiting this kind of movement, and without showing any marked relation between external changes and their own activity. Vegetation, as we commonly understand the term, denotes objects which do not exhibit the kind of movement distinguishing animals. Yet we speak of a tree being alive, with evident implication that its life has something in common with the life of an animal.

The change, then, which is most prominent, and is also most frequent in impressing us, is not displayed by one of the two great divisions of animate nature. Life is not so easily apprehended. Nor should we gain by an attempt *to restrict the meaning* of the word living. The problems would

[1] Nevertheless, the exercise is conditional on a supply of energy from outside the organism.

[2] This change, occurring externally in place and prior in time, is called a *stimulus*.

reappear, and the restrictions would have to be withdrawn.

That special kind of motion which is a marked character of animals cannot be regarded as a sign of all forms of life, since it is not shown by a moiety of living nature. As a test or sign of *life in general* it fails apparently. At the same time, attention may be given here to the fact that movement is never entirely absent from any living organism. Locomotion, that is, change in the position of the organism as a whole, marks the animal; while internal movements, such as may be expected in any body undergoing varied alterations, are alone apparent in plants.

The circumstances of motion give more information than the motion itself. We shall need to touch upon other facts of observation, and to show their range and importance, before we can begin to reconstruct from its foundations a consistent image of life. It is one of the chief difficulties of the subject we are entering upon *that none of its elements is truly isolated.* Each path of inquiry has so many issues, that the most difficult part of our task is to maintain a due regard for proportion. Each investigation has its *own limits* and also its *own special relation* to other parts of a prearranged scheme, and these must be observed. Progress in the elements of a subject cannot always take place along a single path. It has often to proceed by the exercise of reflection in several directions simultaneously.

In our own experience the necessity of taking food is quite as prominent a fact of life as the possibility of spontaneous movement. The periodic satisfaction of hunger is as marked as any circumstance, if we may judge by ourselves; and we may certainly rely upon our own feelings to tell us of the more obvious events concurrent with life. Moving and eating form part of that state, and they are special kinds of activity. All living things exhibit some modes of change. There is nothing, indeed, to bring them under observation but their special modes of change. Then what is the essence of the act of feeding? Is it, standing by itself, a truer sign of physical life than self-movement, or one of wider application?

All animals must take into their bodies certain kinds of matter which are spoken of as foods. This is confirmed by universal experience. It is a fact, too, of *plant life*, equally certain though less prominent. There is some variety in the physical properties of this needful matter, though the differences are *not so real as may appear*. And with regard to this condition, as to many others, our knowledge, that is, the practical, every-day knowledge, is incomplete. According to this, feeding applies to certain solids and liquids taken through a *mouth*. But the real act of feeding, when completely described, must refer to *all* material added to the organism, and this would include the gaseous matter forming the chief food of plants.

Both animals and plants depend for their continued existence upon certain material, which is absorbed and changed in properties by contact with the living body.

Here, then, we have a kind of activity which characterises all forms of life, and it is one to which attention may be paid, as bearing on the statement that living objects exhibit a *specialised form of change*. We must be content, however, with the bare mention of life as connected with activity or change, for the statement can be, at present, little beyond an indefinite combination of words.

We must leave on one side the *details* of the processes which food undergoes within the living organism, just as we must omit to review the *various forms* in which the process is manifested. It may, however, be mentioned that an important character of this process lies in *the exercise of selection on the part of the living object*. It is certain kinds of matter which are alone chosen as food, and we are unable to exclude the idea of selection even in considering the case of plants.

It happens that each of the changes which we have considered so far is noticeably associated with the organism by this mark of spontaneity, of being exercised at will. *Both locomotion and feeding are acts which appear to have as one of their constituent parts an element of choice, and it is this element of choice which is the most distinctive feature of vital change.*

CHAPTER VI

THE OBJECT OF CLASSIFICATION

A Review of living Objects shows the Need of an Order or Plan

A REVIEW of the vast assemblage of familiar living objects, so far as this intellectual feat is possible, brings to light resemblances and differences in outward form. The resemblances may be complete or incomplete, the differences great or small. But varied as form may appear when regarded superficially, a more thorough examination lays bare a still more bewildering diversity. Further resemblances and differences are to be perceived in internal form or *structure*.

The building-up, or *organisation*, of the animal and plant exhibits a diversity in which we may well seek for any hidden uniformity. A procedure which can introduce an *order*, or reveal a *plan* amidst these phenomena, is a necessity of the first importance.

Simple Description of living Forms, however accurate, is inadequate

Up to the present, we have limited ourselves

to mere description. The descriptions given stand as records of observations, which do not necessitate any special skill or training on the part of the observer. The thoughts recorded have no unusual trait, and there is little that is scientific in them beyond their *sequence* and *connection*. Much of the foundation of biological knowledge is of this character, exacting little beyond *accuracy of perception*, such as may be demanded in any mental occupation.

It is important, too, to recognise that practical work in the laboratory, the dissection and detailed examination of objects which have been alive, does not necessarily lead to a grasp of biological principles. Such work, indeed, may end in nothing but *a verification in the concrete of unrelated facts*.

A description of form and structure, however accurate, is not *scientific* of necessity, nor does it become so when the objects described are displayed before us. Something more is needed for our observations to be of assistance in truly enlarging our knowledge of nature, and this comes from the guidance of those who have, in the past, made observations and reflected on them.

An Order or Plan is the Object of Classification

It is clear, then, that description alone is insufficient for our purposes, and the practical realisation of description in the laboratory is likewise inadequate. To be of any value, all descriptive statements must have some reference to an *order* or *scheme* previously conceived. Such schemes have their origin in work already performed. The thoughts and beliefs of an earlier growth must serve as our first guides.

Subservience to a procedure, which has been previously agreed upon, is a first condition of acquiring knowledge, though additions to knowledge may, at a later stage, involve a *contradiction* of the formula by which assistance has been rendered. It is essential, indeed, to the progress of the student that he should make his observations with reference to an idea already existing in the mind of others if not in his own, *and in every scheme of classification such an idea has been formulated.*

The Process of Classification

The process of arranging diverse objects cannot obviously be begun by those who have but a confused conception of the objects to be submitted to the process. Yet the opportunity of placing the forms of life in classes is never entirely wanting, and the necessity for doing so has compelled our attention even from the beginning.

To perceive similarity is to begin to classify. To give the same name, if only to two objects, *creates a class*, and in creating a class we begin the process called *generalisation*. No instrument of thought is used more frequently or applied more universally than generalisation. It is implied in every common name denoting more than one object or quality. It is more or less active, even before names can be articulated.

But we need to get something more than generalisation, whether or not *abstraction* accompanies it, from the effort of classification. A certain order and relation of classes must be a distinctive feature in it. It is true that the perception of similarity carries with it the separation as well as the creation of classes. *But the formation of classes does not necessarily simplify our outlook on nature!* To find an order among objects is the ultimate end of our inquiry. To substitute a confused multitude of classes for an indefinite variety of individuals may certainly diminish the labour of description, but it gives little aid to the understanding.

The Difficulties in applying the Process to living Objects

It is not easy to bring under review all the forms which are now undergoing the changes summed up in the word life. Isolated forms are familiar in their separate existences, though the components which unite to form their manifold diversity are not so easy to discover. Nor does a collection of objects, to which the same name has been given merely on account of similarity of appearance, always form the kind of class to enlarge our knowledge of Nature. A process of this kind, without ulterior aims, forms part of the mental activity of every individual at every stage of experience. *But a classification to be of service in Biology, or in any science, must refer to something beyond the superficial appearance at which ordinary inquiry stops short.*

With the actual results which have been obtained from classification in this science, we are, for the present, less concerned than we are with the mental creation, which precedes it and gives to it all its force. The centre of interest lies in the ideal framework which assists in the separation and distinction of phenomena.

In Biology, *the living organism* needs to become a definite conception for us. It is an abstraction which is as difficult as it is important to achieve; and any arrangement exhibiting an order of concrete examples must display this abstraction as its chief content. On the other hand, even a crude classification may give valuable assistance *in the shortening of statement and the simplifying of communication by language,* and any gain in conciseness of expression cannot fail to extend the range of our thoughts. We cannot, therefore, afford to neglect such an instrument of progress at any stage of our course, however imperfectly we may be trained in its use.

If the larger aims of classification are kept in mind, the characters of biological classes should be disclosed progressively. The full meaning of such a scheme is not at once apparent. Our comprehension does not always move onwards by easy stages. The mind often needs to make repeated excursions from *some central idea*, and that not necessarily a very clear one, if it is to succeed in enlarging its circle. *The organism is the conception at the centre of all biological excursions, whether they take the direction of classification, history, or ontology.*

The dual Aspect of the Organism as possessing Form and also as presenting Change

In any arrangement of living objects it will be necessary to pay attention to the two-sidedness which they manifest. The *activity* of the organism must be regarded as well as the *form*. At the same time it must be admitted, that the form is most readily apprehended in connection with its specific activity. Or we may go below the surface, and consider the definite *group of structures*, which plainly constitutes that form in so many instances, as the seat of a definite *group of recurrent changes* jointly conferring the quality of life upon the whole substance.

It is, indeed, not possible to think of the changes apart from the objects undergoing them. This is true, even of the simplest kinds of change.

Displacement, for example, is a change in the position of a body. But there is no such thing as position without *a body to mark it.* Nor again is time intelligible without material objects of some kind to define it. With the far more involved changes of alimentation, sensation, and the like, the implied reference to an organism is obvious. Time, space, matter, and life are all relative as well as mutually dependent ideas.

And this is the real meaning of the word organism—*the something in which life appears, the substance which is organised for life, the body to which life must always be referred through the limitations of our knowledge.* Hence an arrangement of all the many kinds of organisms under a coherent scheme only becomes possible when we begin to understand how activities are connected with, and conditioned by, *the forms and structures which are organic to them.*

A joint Regard to Form and Function simplifies the Work of Classification

Variety of form, then, is manifestly too indefinite and too meaningless to be taken by itself as a basis of classification. An arrangement of a certain value may be founded upon it, and, indeed, classes based upon observed similarities of appearance have always served for the ordinary wants of mankind. But the ordinary wants have been nothing beyond directness of denotation and economy of description. Further wants arise with the growth of inquiry, and the effort to attain to an ampler generalisation which shall express the results of wider observation, is the beginning of a scientific system. *The search for "causes," or correspondences between facts, is a development which is logically subsequent to the desire to describe with brevity.*

A scientific scheme, with comprehension as its aim, takes the place of the simpler and less consistent results of generalisations, based upon appearances which are more or less detached. The results in the latter case are disconnected, or only accidentally connected, while a due regard to function brings about a systematic connection among classes. All classes are alike with respect to certain kinds of change (though the circumstances of the changes may vary widely), and a relationship between them is instituted as soon as the attributes in common can be clearly perceived.

There is *a unity* underneath the diversity of form, which the intelligence must set itself to discover. It will be found in a correspondence of action or function. Now functions are limited in *kind*. They are primarily such as relate to *assimilation, sensation,* or *reproduction.* Other functions exist, it is true, but they are secondary in some degree to these fundamental activities.

The function of a given structure, that is, *the habitual activity which is most prominently characteristic*,[1] when considered in connection with the structure or organ itself, gives rise in our minds to an image which is more easily apprehended than either function or structure by itself. Indeed, the mental isolation of the activity from the object exhibiting it is often a difficult process. *This statement may be made of the organism as a whole, just as truly as of parts of the organism.*

Our perception seizes most readily upon *something undergoing change.* The distinction of form from activity, that is, of Morphology as apart from Physiology, is an achievement in abstraction to which we attain at a relatively late stage in the pursuit of knowledge.

The need of food, which is common to all forms, gives rise to a possible basis of classification. Since all organisms assimilate certain matter, herein is a condition of life by which all forms are brought into relationship. If it be now desired to make classes, and to compare one with the other, the variations in this common activity may serve. And it is easy to gather *that variations will be most perceptible in the difference of structure which accompanies them.* A difference of form marks a difference in the *mode* of the function. The *initial* and *final states* are approximately the same for all. The intermediate stages exhibit variety, and give occasion for

[1] See p. 98.

classification. The events contributory to assimilation, as manifested in structural variation, are crowded enough to indicate not only resemblances and differences, but degrees of the same; and it is *the perception of distinctions in difference which renders a classification systematic, and makes it an orderly arrangement.*

A similar procedure may be adopted for other of the universal attributes of life, *e.g. reproductivity* or *excitability.* They are properties possessed by all, but exhibited in various modes and degrees. *We may place together those organisms which are alike in their relation to a common character; and, by reason of distinctions in relation, we may give to our classes an order or place in a scheme.*

No systematic classification, however, has been based upon the facts of assimilation. Reproduction has been more serviceable, especially in the classification of plants; while the prominent fact of excitability, which gains fuller meaning in the words, *the relation of the organism to outer changes,* has not yet been utilised. *On the other hand, a perception of all three characters, as existing in combination, has given rise to an idea called development, and this has afforded a ground of classification which has proved the most trustworthy up to the present.*

The Idea of Development in Classification

The word *development,* which has a common

usage in Biology, may denote either a *process*, the act of developing, or a *state*, the result of a process. The present meaning is restricted to the state or condition which arises from a process of change. It is not easy to describe what is meant by a state of development. There is obviously something, which varies in degree and may reach different stages in different objects, but the something needs to be indicated.

In every living object, even the simplest, there is some appearance of structure. In other words, the several parts do not present the same appearance to the observer. What the cause may be, whether the substance of one part is truly unlike that of another, are questions needing an answer. There is always some indication of structure and organisation in the material of life. An estimate of its complexity, that is, the extent to which a difference in parts is apparent, may be attempted by the ordinary methods of comparison.

But substances which are bearing the stress of life seldom disclose their structure to the first glance, even if the difficulties of inspection be overcome, nor do they often admit of comparison. An equal degree of dissection or analysis would be required for each, and *as mere size is of no consequence in questions of organisation or complexity*, it is essential to be in possession of some standard before it is possible to institute a comparison. *The need of a standard for the purpose of estimating any value or magnitude, whatever its kind, is an axiom of science, nor is the highly composite and abstract magnitude of organisation exempt from this necessity.*

It is neither advisable nor easy to judge by appearances. The enumeration of differences is a help, the greater number corresponding with a higher degree of complexity. But this is not enough in itself. Our conception of the organism is a dual one, structure together with function being the component ideas. These two aspects of the organism must be considered together. The true problem before us in estimating development is to understand how function and organ are joint factors in the degree of organisation. It may be said that structure is meaningless except as related to function, and function is inconceivable apart from the medium in which it is manifested. *A separate manifestation of individual functions cannot be shown except through variety of material structure; in one sense, structure and function are the same fact.* The modes of expression here differ more than the things expressed.

There is one feature which stands out in marked prominence whenever we consider the correspondence between structure and function. The combinations which are possible in structure are more varied than is the case with function.

The kinds of function, as has been previously stated, *are limited, while the modes of organisation which are contributory to them are most varied.* A real correspondence is first perceived to exist when we become aware that functional *processes* are more numerous than functional *ends.* It is the variation in the mode of securing certain ends which brings about different details of structure.

A fairly accurate conception of the organism, as that which displays itself by a certain minimum of functional changes in a complex substance, may be obtained from these considerations. Divergences between organisms are manifested primarily as differences of material complexity, and only in the second place as functional differences. *The development of an organism is its state with regard to both of these complexities, and it is estimated by comparison with the simplest object which is observed to manifest life.* Such an object, whether real or imaginary, might serve as the standard of which the want has been described.

Yet the ideas here represented are not equal to the purpose in view. A scheme based on them would not give information either correctly or adequately. The extent of development is found to be, when taken by itself, insufficient for the task of classification. The degree of divergence from a single standard form, *i.e.* the number of resemblances or contrasts, does not give us an order in the relations between classes which is an essential feature in systematic classification.

The Conception of Types

If the facts of life were simpler, if life did not appear to us in every situation with such varied modifications of its known processes, and with a scope of form to match, the creation of a system would be a comparatively easy matter. The variations in complexity, whether of structure or functional processes, might have taken the form of a *simple linear progress*, each successive increment of complexity giving rise to a successively higher rank in the scale of organisation. But however important it may be to regard development as the ultimate basis of classification, we cannot perceive in the whole body of organisms any distinct signs of such a simple progression. Nor is it easy, on the other hand, when we regard a single individual, *to separate, even approximately, any one function completely from neighbouring events, or to succeed in accurately isolating or localising any single class of activity out of the many which combine to constitute its unitary existence*. If living changes could be more readily contemplated in their entirety, and each be traced by itself through an undisturbed course, results could then be weighed more directly and most of our difficulties would disappear for good.

The mode of complexity may vary as well as the extent. In attempting to classify organisms some modification of the ordinary estimate of aggregates is obviously needed. The quantitative method fails, while a comparison with a series of distinct schemes of structure affords us the first insight into the general plan of living Nature. *Hence arises the creation of certain ideal combinations of characters called types, which serve as a series of separate standards of reference.*

A comparative review of either animals or plants shows that *no single character* is of use as a means of classification, nor indeed can any *simple combination of characters* serve that purpose. The simplest combination upon which a comprehensive and intelligible arrangement has been founded is illustrated in botany. We find systematic botany mainly consisting of a scheme of classification of plants which is based upon the structure of flowers, the form, number, and arrangement of the floral whorls serving as a guide in the separation or grouping together of the objects of which they form part. On account of the prominence of the flower as a part of organisation, and the importance of its function, there is much to be said for this method, and the results are of undoubted value.

When the far more divergent complexity of animals is considered, a reference to types appears to be the only satisfactory means of diminishing the confusion of our impressions. The meanings which are assigned to the word *type* as used in Biology do not always agree. There will be no error in regarding a type as *an ideal construction derived from a number of real impressions*. From observations of similar objects, or objects with more than one quality in common, we are able to create in our minds *a conception of certain combinations of qualities as existing in a number of individuals*. The type has no real existence by itself. It is a conception of a definite group of relations, presented by objects which are alike as regards the relations composing the type, though different in other respects. One group of relations enters into the composition of the *mammalian type*, another group into that of the *echinoderm type*, and so on.

By a full recognition of the ideal nature of types, an effective comparison of objects with one another is rendered possible. *There are as many types as there are distinct groups of qualities, and as many classes as there are distinct breaks in the sequence of development.* The discrimination of distinctness in this connection, it is true, may vary, but there is, at least, no question as to the gain in order resulting from this method.

And though important problems arise as to the relations to be instituted between these *groups of relations*, the useful and orderly arrangement of living objects achieved on these lines is quite independent of all assumption and theory as to the origin of such relations. A classification has been eventually achieved without admitting any speculation as to the *ultimate cause* or *origin* of classes, though speculation may be quickly aroused when the abundant evidence of relationship between classes is first clearly discerned.

The Influence of Authority

It is an accepted fact that there can be no progress in the acquisition of knowledge, without a constant reference to the history of the subject and to contemporary opinion. Yet it is frequently found that controversy is at once aroused by any term which happens to embody a recognition of opinions already formulated. It would be difficult indeed, in any course of inquiry, to rely solely on personal observation; yet it is often assumed that there is something antagonistic to progress, and hence in contradiction to the true scientific method, in that dependence on others which is implied by the word *authority*. But it is not necessary to demonstrate that which is obvious, the futility of any investigation undertaken without some sort of guidance; nor is it easy to imagine any epoch in our mental career when our thought has been quite free from external control.

To be independent of past and ignorant of present achievement, to disregard the accumulated experience of all who have observed, considered, and believed, is not a satisfactory state in which to undertake any systematic research. The unassisted intelligence is a poor tool, and the total experience of an individual is a slender fraction of the whole knowledge which mankind has amassed. Yet the larger knowledge is sealed from those who are unable to follow the methods by which it has grown to such enormous dimensions. To give a training in those methods is the chief end of education. By employing our intelligence in following the methods of others we make their knowledge our own.

Any formulated thought is *authoritative,* in so far as it represents the collective result of observation and perception which has stood the test of time and controversy. And Biology is not a branch of science which can afford to neglect the generalisations which stand as records of past researches. The science of living things covers too large a field of survey to be successfully explored without guidance. It is in the earlier stages, however, that authority is most valuable and most appropriately relied on. Once the observation has been trained and the intelligence formed, the individual may stand alone; and by his own efforts may hand on, in his turn, some contribution to the common stock of knowledge from which his successors may draw.

There is no escape, then, from the influence of authority in scientific matters, and this must be duly recognised. An authoritative statement expresses the *highest point of positive comprehension* to which the intellect has hitherto attained. It shows the position which thought has taken up after controversy has rendered it exact.

Knowledge grows more exact by Communication

On taking the trouble to consider this quality of exactness or precision in current thought, we shall find, that so far as it exists at all, it is largely dependent upon the agency of language, by which thoughts are communicated from one person to another. The words which act as symbols of subjective states do, indeed, in their use, help to render those states more precise. A spoken or written word is a definite sign, and a feeling or a thought is frequently vague or confused. *The ordinary use of language*[1] *tends, therefore, to increase the determinateness of thought.* And this is because a given term is a sign, not of the whole confused thought, but of that in it *which is capable of taking a clearer outline.* It may not be going too far to say that science begins as soon as ideas are exchanged, for the exchange involves a process by which the ideas acquire more definite form.

All communicable knowledge is therefore to be looked upon as more or less exact. It would

---

[1] And the use of diagrams or any system of signs for the purpose of description.

not be communicable if it were not exact, and the process of communication tends to intensify any sharpness of outline already possessed. There will be no failure, it is hoped, in perceiving that the term used in this connection is exactness and not accuracy. It is not intended to imply that any transcription of ideas into words can make them infallible. *On the contrary, many exact opinions are erroneous, but it is their quality of exactness which brings them within the scope of inquiry and discussion.*

Opinions of Others must be recognised

The records of previous observers must serve as guides to our own work. We are not always certain of truth, though we are judges of what is inconsistent with opinions already formulated. Propositions which contradict one another, however exact in expression they may be when standing alone, cannot each be true. We are, however, able to make a safe start on any inquiry when once we are furnished with some of the great generalisations which have received the sanction of many thinkers, proved useful aids to the extension of knowledge, and assumed in the progress of time clearer and likewise fuller meaning. Herein is again apparent our dependence upon authority.

But it has already been stated that a complete isolation of the individual worker is never attainable. He is never free from obligation to pre-existing opinions, upon which his own work is necessarily founded. At the outset of this inquiry, words were used which assumed the existence of mind; for *thought, idea,* and *knowledge* are terms which imply in their use an advanced capacity for abstraction. And we shall find at all stages many things to be assumed and taken for granted. There is nothing gained perhaps by dwelling on this state of dependence. It occurs too frequently for us to remain long in ignorance of it. To search too patiently for the reality of those appearances which have, admittedly, to be regarded as obvious or elementary is not conducive to progress. Again and again we have to proceed, even if we feel that we are being guided past serious difficulties. We may console ourselves by returning to them at a later stage.

The significance of authority is never better understood than when one attempts to trace the origin of a system of classification. It has been pointed out that the mental effort to arrange our complex world of objects in some order is the basis of any scheme able to justify its existence. But the system with which we have to be content, in fact, our conception of Nature as a whole, is largely derived from the experience and the mental exercise of others. *The growth of intelligence, however spontaneous it* may appear to the individual, is in reality under incessant control.

Our classification is initiated by authority, and it is based upon assumptions which we are, at first, unable to revise, even if we succeed in understanding them. In truth, every scheme for the objective world which arises in the individual mind has a second-hand basis, whatever may be the originality of the superstructure. If we keep this important truth prominently before us, the real value of our classification will be raised rather than lowered. *The mental scheme or belief, as we may call it, will not take a lower rank on account of a recognition of what is a condition of all progressive knowledge or science.*

Summary

There is, then, one requisite of classification which must never be lost to view. There exists a circumstance in the act of arrangement which at once distinguishes it from an aimless activity of the senses. It is a condition which is admitted to exist before the process of arrangement begins. No classification is capable of aiding our comprehension of Nature, except through an independently existing plan, for without an object or end in the provident mind it ceases to be classification. This holds true, whether the classification is regarded as a disposition of objects in space, *a real grouping together* of similar things: or *a mental vision of such a disposition*, based upon some experience of such an act: or, lastly, whether classification is *wholly an ideal process*, one, in fact, which consists in an arrangement of our ideas rather than of objects.

The primary condition of success in classification is overcome as soon as we perceive that the mental process of conferring an arrangement or order upon objects must go on during the exact observation of those objects, even if it do not give rise to that exactness of observation. In other words, *we must classify while we observe, and this statement is true of other sciences besides that which deals with the forms of life.*

It will be urged by some that our experience does not exactly coincide with this statement, and that observation precedes arrangement, although arrangement or classification may provide opportunities for extended observation. It may be, indeed, that the truth lies mid-way, and each process reacts to extend the other; but that which is most likely to be lost to sight is the operation of the principle just expressed. Even if some of us remain uncertain as to which has the greater share in the intellectual value of the result, we may admit at least that both are active in assistance.

Nothing is more valuable than the simplification of thought which must necessarily follow the gradual inclusion of all the visible perplexity of form and organisation in a comprehensive system of ordered and related images. And that mind is best fitted for investigations into Nature which is already endowed with some conception of order. *It is indeed the order conferred upon Nature by ourselves which makes it an object of systematic and intelligent investigation.* The belief in an order or plan is the most important equipment for inquiry. Nature cannot be regarded reasonably as a confused multitude of individual facts, existing outside ourselves and waiting to be observed by the students of special sciences.

CHAPTER VII

CERTAIN GENERAL STATEMENTS CONCERNING
ORGANISMS AS INDIVIDUALS

Explanatory

THE heading given to this section demands a few words of explanation. It indicates an intention to furnish the student with further aspects of the characters and changes generally observed in living objects. The statements to be made are not necessarily of characters or changes hitherto undescribed, nor are they less important for that reason. A single act of observation does not confer much intellectual advantage, unless its results be compared with previous experience, as occurs in the process of reflection. An impression once gained by the observer needs to be recalled repeatedly, for purposes of comparison with others of the same kind, before it can rank as a part of knowledge. And some of the facts here described are not new. Their application, however, is in most cases modified. The same things are looked at from a changed point of view, and new relations are consequently disclosed.

In addition, it may be explained that the phrase, *a general statement*, is one which is supposed to be universally applicable. It is made not only of one object, but of all. Since it expresses the result of innumerable observations of the same kind, being a sign of similar impressions frequently repeated, it is believed to be true in general. Yet it is important not to lose sight of that which is taken for granted in a general statement. Assumptions do not grow smaller with repetition, though they may escape our notice more completely.

Such as they are, the statements are made about *individual organisms*, that is, living objects which are separate and distinct in space. And they are made about organisms in their capacity as individuals. We have to learn to regard the individual as something which is complete in itself, and can disclose in any of its manifold activities a single end, and this end is its own existence as an organism. A conception of this degree of complexity is not easily gained. And the other side of that *objective isolation*, namely, a *separateness of feeling*, of which we are conscious ourselves, must greatly complicate the outlook if extended to every instance of life.

Life is continuous, Forms are intermittent

There is no difficulty in recognising the truth of this general statement. On the surface of the

82 THE LIVING ORGANISM CHAP.

earth life is manifestly continuous. There is no observation, nor any inference to be drawn from observation, which can give the slightest indication of any *general cessation*, if only temporary, of the changes associated with life. It is inconceivable that the total sum of activity, to which every living item contributes a share, should cease to exist for a given period and then begin anew. Such an event would be outside the course of Nature and contradictory to all experience. There has been from the beginning, somewhere and in some form, an unbroken sequence of life.

In contrast with this statement of fact we may place a general truth which is equally important, the finite duration of the individual living object. With regard to any given organism, the changes we are able to observe have their limits in time. That the *individual* is mortal is as patent a fact as the *general continuity* of the kind of change known as life. The characteristic changes in a given object which are regarded as signs of life will, sooner or later, cease as a whole. They do not occur for more than a limited time in connection with the same form or identity. The forms of life are intermittent, for the individual dies. *The changes which signify life persist.*

Reproduction an Event which ultimately brings about a Disturbance of Form without a Break of Life

The repetition of the forms which manifest

life proceeds from a constant character in the organism, and the *periodic disappearance and reappearance of forms* is as distinctive of living matter as anything which has yet been noticed. Indeed, the phenomenon of reproduction divides living from inert matter no less clearly than the more material event of nutrition.

It is observable that the organism, whether plant or animal, may separate from itself a portion of its living matter, and this fragment may then lead an independent existence. In other words, it will answer to any of the tests which we may choose to apply in quest of its completeness and unity. It is a distinct living organism no less than the parent from which it was derived. There is, so far, a consequent increase in the number of individuals.

Now the most general impression to be gathered from the ordinary process of growth is that of an increase in the quantity of living material. At the same time, the change which has been called reproduction may be looked on as a kind of *discontinuous growth*, for its first and obvious effect is an increase in the *number of individuals*, as well as an increase in the *quantity of living matter*, except so far as other changes diminish that quantity. It is *more* than growth, for that is an increase in quantity alone. *There is in the reproduction of the individual an event of a distinct order.* The first steps in its comprehension cannot be taken until the meaning of individual is better known.

Death is a natural Sequel to Reproduction

The event to which the term *death* has been applied is not to be regarded as a simple occurrence. It must not be looked on as similar, for example, to the stopping of a body in motion, nor must it be regarded as the instantaneous cessation of all those changes which constitute life. And we do not gain the fullest knowledge while considering it as a gradual diminution of activities, occupying a longer or shorter period of time, although this view approaches more nearly to accuracy. When the activities which constitute life are better discerned, and the existence of an individual as such is more clearly understood, the conception of death will be more exact, for it will then be regarded *as a break in the relation or co-ordination of vital changes, not as an abrupt termination of all activity.*

But at the present moment we are not so much concerned with the process of death or its nature, as with the relation of that phenomenon to the other facts of life. Death comes to keep the balance between conflicting operations in Nature. Unchecked growth and reproduction would militate against the activity of life. The material conditions of life are such that its extent or quantity must be limited, and death is the natural limit. *The Individual, Reproduction, and Death constitute the tripartite scheme of organic Nature which contains and dominates all else.* These three component phenomena are *periodic, mutually dependent, self-adjusting,* and *universal.*

A further Discussion of Growth

The process of growth is displayed primarily as an increase in the matter or bulk of the object under observation. It is possible to learn more about growth, by keeping the attention more closely fixed on the individual. Yet we have to admit that this is not easy. The individual may be readily described as to position, and as to duration of time, but with regard to the matter composing it no complete definition is possible.

The periodic absorption of food is enough to show that there is no constancy in the material components of the organism, even if the obvious facts of waste did not point in the same direction. It is true that there may be no change in the total quantity of matter composing the individual, while on the other hand there may be either a diminution or an increase of quantity.

We must still be a long way from understanding the organism and its growth, if we can only assert that it is an object which exhibits certain changes for an indefinite period of time, probably in vary-

ing localities ; while identity as regards matter can never be asserted, and, indeed, is safely denied.

An object which is not the same material object for any two consecutive periods of time, does not impress us as safe ground upon which to come to conclusions about the change called growth. So little is fixed and permanent in the material of an active organism, *in spite of the absence of any sudden break*, that we are frequently driven to disregard the material and make our own image of a something through which a stream of material is constantly passing. Shadowy as such an image may be, we need its assistance in many of our inquiries.

But, without dwelling on our ignorance of all the changes suffered by matter while it forms the substance of an organism, we must pass to the evidence of such stable qualities as do exist in that substance. Discursive as the previous expressions have been, they may succeed, at any rate, in indicating how difficult it is to make definite statements about growth.

The idea of *assimilation* is one which may now be suitably introduced. The word implies a certain capacity on the part of the organism to affect matter brought into contact with it; and it is used when we wish to draw attention to the *process* rather than the result of the event of growth. It furnishes, too, another illustration of the slenderness of our positive knowledge.

Assimilation is the Reality underlying Growth. The Ability of living Matter to assimilate other Matter is to be regarded

We are unable to trace any given portion of food to any particular locality within the organism. Many chemical and physical events ensue on the first contact of food with the organism, but these are clearly · preliminary to the act of assimilation. At what point in its course the food enters the inner region of living events is hard to say. It is certainly impossible to fix assimilation at any single instant or the endowment of life at any given place.[1]

Yet during a temporary and unlocalised sojourn within the body, each particle, so long as it is truly within it, does emphatically form part of the body. It has been assimilated, or made like the neighbouring material which was already resident in the body, though only temporarily resident.

The question of total quantity, whether increasing or diminishing, becomes relatively unimportant. What is noteworthy is the act of assimilation. And this requires among other things that the ingested material shall be capable of being assimilated. It must resemble in some

---

[1] 1. Is the matter alive only *when itself assimilated*?
2. Is it alive only *when it is assimilating other material*?
3. But does not the state of 2 in itself imply that no portion of matter can for long be alive, certainly not coincidently with the life of the organism?

degree the object with which it is about to be so intimately allied.

Here too the exercise of choice on the part of the organism has to be noted as a prominent part of the process. *The something, through which material is passing, is capable of selecting that material, either actively or passively.*

What is it that assimilates?

Yet the something, capable of selecting, changes while it selects. It is certainly not the same matter which continues to exercise assimilation. *The individual is continuously alive, but its identity does not depend on identity of material.* It is perhaps not wise to speak of the individual as an activity or a capacity, though these words may suggest themselves without, it may be, any accompaniment of exact thought. We might in such case, indeed, be compelled to speak of the organism as a "*capacity to assimilate.*"

Without doubt the most exact impression made upon our minds, while the organism is under consideration, is reproduced in the idea of form. *The matter changes, but the form remains;* and it is the constancy, relative only though it may be, of the form which is the most fixed sign of continued existence.

Reproduction connected in some Way with Growth

There is great variety in the conditions under

which reproduction occurs. In some forms, and they are, as a rule, those of more complicated organisation, the detached fragment is a slight fraction of the whole. In some of the simplest forms, on the contrary, the parent organism divides more or less evenly into two individuals; while in others there is a more or less complete disruption of the parent into a varying number of distinct individuals. There are also occurrences which are intermediate in character.

But there is an important distinction to be drawn between two modes of reproduction, which are alike in their result. On the one hand there is the process of outgrowth from, or disintegration of, a single individual; while on the other there is a fusion of portions of two separate individuals, before the development of the new generation can take place.

However varied may be the modes of the latter procedure, that which is essential to it is a conjunction of material from two individuals. It may be, indeed, that there is a coalescence of material precedent to every occurrence of reproduction, though the necessity of such an event is not distinctly understood. But we may see in every case of reproduction a connection with the process of growth, whether it appear as a temporary increase in its rate or as a multiplication of the centres of assimilation.

It is difficult to omit, in considering this

problem, all reference to the subsequent history of the detached germ. There are two most striking features in the process of reproduction, whether it be an instance of the more simple *agamogenesis* from a solitary individual, or an action in which two individuals are concerned (and therefore distinguished by the term *gamogenesis*). In the first place, there is exceptionally rapid growth as a whole. In the second, there is a rapid development (or differential growth) from the comparatively structureless ovum to the organisation of the mature individual. *The habit of growth is transferred from parent to offspring.* Not merely the capacity of formless growth, but the capacity to grow in the special manner which has distinguished the parent, has been handed on. The new organism is not merely alive, it exhibits the special kind of life which is shown by its progenitors, their functions and structure being ultimately repeated.

The Association of Death with Reproduction

What has now been described is the most general view of a fundamental character of living matter. It may be expressed shortly by saying that life is *periodic* in appearance. *The isolation in space*, which is the consequence of reproduction, is compensated by a *sequence in time*, which is the essence of an event invariably associated with reproduction, namely, death.

As far as our observation extends, the individual from which the second individual is detached, or, as we term it, the elder individual, is more prone to death. In the normal course of events the young succeed the old, which they grow to resemble, and so on indefinitely. The "lamp of life" is handed on from one generation to another, and the individuals related in the special and mysterious manner indicated by the word generation present, as a rule, similar forms.

The Permanence of the Form

We may take it, then, that the existence of isolated and separate forms of life, persisting in spite of the recurrent phenomenon of death, need not engage our attention any more than the less familiar or less closely scrutinised fact, of the same character or impress being given to the *succession of organisms* which are materially related as regards life.

The repetition, or reproduction, of similar forms makes considerable demands upon our comprehension, far more indeed than a possible process of separation or detachment of a portion of the living matter from the parent organism. This, it is plain, we might regard appropriately as an *act of superabundant growth,* and the *isolation in space* as an accident due to the physical conditions of growth. It may appear to some that the form is impressed from without, that the living and growing matter, being pliant, is moulded by unobserved agencies external to it; while many will be satisfied with considering the "cause" of the expressed or impressed characters to be resident within the organism itself.

Now the material aspect of the process ending in the attainment of form is clear, for it has much in common with all the material changes connected with life. It is the *sequence of individuals*, the so-called death and the so-called birth in alternation, that contains the mystery. Lest there should be a further misunderstanding in this connection, it may be as well to state emphatically that no biological explanation of life and death is possible. There is no analogy to be drawn from other changes in Nature. *The recurrence of the individual is a unique phenomenon.* The material concerned in it may be observed and watched, the facts may be generalised, and the whole may be philosophically contemplated, but there is no explanation nor comprehension of it to be gained.

Summary

The terms which express the subject of thought in the last few pages, *growth, individual, reproduction, death,* and *life* itself, stand for a connected body of ideas which have been discussed merely in outline. They are ideas which should be, as far as possible, considered together, since they are all related. It is not too much to say that they have something in common, inasmuch as thought about any one of them leads insensibly to thought about others. Perhaps it may not be out of place to suggest that it is a perception of unity or isolation, of oneness or separateness, or whatever we may call it, which is the common ingredient, and the main ingredient, of all these ideas.

But thoughts about any one of these subjects will always bring ideas of the others in their train. Yet a very slight change in the course of thought, starting from the same primitive conception, may sometimes bring about very considerable differences in the conclusion. The term *death*, as denoting the cessation of the observable changes which characterise life, has a merely local application. The thoughts in connection with it cannot be separated from the idea of the individual, just as its negative, *life*, can only be asserted of this or that material object.

It must be remembered, too, that death and life are the names of *occurrences objectively viewed*. No one of us has felt death, and no one can experience the life of another person, much less the life of an animal or tree. In both cases we are discussing certain external events *as they appear to us*, and each event is composed of many appearances. Death is a complex group of changes, and life is a manifold occurrence in a definite object.

The object of which either of these events may be asserted is an organism. The term individual, not a wholly satisfactory one, usually has the same meaning as organism. But putting on one side the question of distinction between them, the terms act as symbols of nothing apart from consideration of life, or the possibility of death. There is, therefore, the thought of oneness or separateness always brought to the mind, if it be rightly impressed by the idea of the organism. And although we may talk about life in general, we do not know of any such thing except as the special phenomenon of a definite organism. Life clearly has unity, or individuality, at the core of its meaning.

Passing from the contemplation of the organism as an individual presenting its own group of changes, more or less completely independent of the existence of life in other subjects,' we are confronted by the fact of reproduction, and it is this event which co-operates with death in redistributing the forms of life.

Though it is essential to regard the individual as an object, with boundaries in space which are distinct for the time being, yet it is clear that the effects of assimilation may consist in some modification of size if not of shape. We have afterwards to familiarise ourselves with the idea of the occasional separation—an isolation in space as it is—of a portion of an organism, without any loss of those properties which we recognise as vital.

The abnormal division of an individual by an external agent may result, in some organisms, in each or every portion retaining its vitality, and also its ability to pursue a separate existence and to grow eventually to resemble the original. But in reproduction we perceive a normal part of the scheme of Nature, a spontaneous fragmentation, if the expression may be used, of the process of assimilation. Instead of one centre or locality of vital changes we have more than one. The material continuity of the organism has been interrupted, and the individuality either multiplied or, perhaps, divided.

But again are we met by a difficulty previously encountered. How do we know, or how can we describe, a form apart from the material of which it is an observable quality; and how can we picture to ourselves the disjunction of that imperfectly comprehended form into separate individuals, as it occurs in reproduction?

We find once more that each aspect of life involves the recognition of others, and when we come to put our thoughts into words, we find that there is no serviceable knowledge to be derived from isolated impressions. We are driven to the conclusion that the ideas which help to make up a conception of life are dependent one on the other, that the terms, *form, organism, function, growth, assimilation,* and others, stand for an interwoven network of thought, which must be treated together, and yet also in part discriminated.

CHAPTER VIII.

A GENERAL DESCRIPTION OF THE ORGANISM AS RELATED TO ITS SURROUNDINGS

A short Review

WE have now reached a point from which it will be a great advantage to retrace the steps already taken, before attempting further progress. And a review is all the more necessary on account of the many difficulties which are sure to be encountered in the mode of treatment adopted in these pages.

So far the attention of the observer has been directed to certain modes of activity as characteristic of life, and to the existence of diversity in the forms which manifest life. We have taken it for granted that our direct observation of living objects comes under two headings—*form* and *change*. Inquiry into the various forms, and into the structure and organisation implied in those various forms, make up one branch of our research; while the various modes of special and distinctive activity, or function[1] as we may call it, constitute another region for investigation. To give priority to one or the other of these subjects would be meaningless. They rank together, and cannot well be studied apart.

The Relation of Subject and Object

But there is another matter to be considered, which is an essential part of all exact knowledge, and one which in connection with Biology is especially significant. It is difficult to describe this important branch of our work otherwise than as the relation of the observer to the object observed. Care and attention must be given to its treatment.

Already the three terms, *observer*, *object*, and *observation* have been frequently used. They have now to be considered in connection with each other, or rather, the ideas which they represent are to be brought up together for review. In the first place, we may say that the *act* of observation is a connecting link between the observer and that

[1] Function is the term used to denote the quality or qualities in an organ or part of an organ by which it is enabled to contribute to the activity of the organism as a whole. The term function does not apply to any activity in a portion of living matter, nor does it always apply to the one which is most prominent. It always refers to a special and characteristic change which gives, in conformity and in combination with others, a complex resultant activity known to us as the life of the organism. A purpose or end is always to be associated with a functional change.

which is observed, the object. Or, perhaps, it is better to say that the observer and the object are brought into relation by an activity, which has its origin apparently in the observer alone.

This activity is a real event, and it is a most striking character of some instances of life. Yet it is not one of the changes already described as characteristic. It is not an observable change at all. But it stands for a whole group of important facts, which have to be included in the conception of a sentient organism.

The relation between the observer and the object, as expressed by the word "observation," is the same as that which is implied in the use of the words "subject" and "object." And these words carry in their meaning a distinction between *feelings* and the *material* or *causes of feelings*. Feeling is an important part of the life of the subject, and so long as feeling exists an environment of objects is a necessity. It may be taken then for granted, that this fact of life with its *two aspects*, the outer appearance and the inner feeling, must be scrutinised before attempting to generalise about the whole.

The Objects of Study are like Ourselves within certain Limits

We ourselves possess those attributes which form the material of biological inquiry. We investigate the manifestations of life by means of some of the activities which make up our own life. Ideas derived from our own feelings of being alive, our experience, for example, of movement and hunger, must form part of our thoughts about other living things.

Do we, then, observe from *within* or *without* the system of activity which we call Nature?—is a question which naturally arises. Without attempting to answer it fully, it may be said that no scheme of Biology can be complete unless it include some regard for the living intelligence which observes, reflects, and constructs the science. A consideration of the forms and functions of animals must, therefore, be supplemented by some attempt at *an analysis of the process of thought*, before a satisfactory and adequate conception of life can be gained.

At the outset we used terms which rendered the latter statement inevitable. *Thoughts, ideas, conceptions, knowledge,* are terms which assume a very important part of our own activity. We ourselves are alive, but our thoughts about our own life require a mode of expression which we are not justified in using about living things unlike ourselves. Ideas, conceptions, knowledge, and such words, do not readily suggest themselves in connection with many things that are truly alive. There may be *action* or *doing* in common to all, but there is a whole world of feeling of which we know nothing, except by inference founded upon our own experience. When I consider myself as alive, I find I can sum up my bare existence, without entering into questions beyond the scope of scientific inquiry, under the three main headings—1. *I feel*; 2. *I know*, and 3. *I act or do*. To how many living things do these statements apply?

We differ from other Organisms in Part of our Relation to our Surroundings

At any moment of life we are in one or other of the states expressed by these three phrases. *Feeling, knowing,* or *doing* may be asserted of us so long as we live. Yet we are unable to assert more than one of these states, that of *doing*, of most forms of life. By an extension of the meaning of the word *feeling*, that also may be held to denote a condition in the existence of many, if not all, animals, and possibly also of plants. With a still wider meaning understood, it is not an easy task to fix limits to the existence of feeling. It becomes difficult to draw the line between inanimate reaction and organic feeling.

But it is not with the intention of presenting a contrast between ourselves and other organisms as regards physical life, that the above phases of our existence are quoted. It is rather with a view of arriving at the basis of our own relation to the rest of the world, and especially to that living part of the world about which we are trying to form judgments. We are not much concerned at present with the feeling or the knowing of animals and plants, for there is no image which can be brought up in our minds to correspond with such a juxtaposition of terms. We have no grounds at this stage for possessing ideas on the subject. But it is quite certain that one or more of these states, which are a true part of life in its fullest expression, furnish the opportunities whereby we learn, and form judgments, about that part of the world which we consider to possess the attribute of life in common with ourselves.

We cannot afford to neglect a discussion of these matters, and, as has already been pointed out, something of the processes of thought must be learnt, and the modes whereby the observer is enabled to observe must be sketched, before knowledge which is permanently satisfactory can be attained. In fine, we ought first to deal with the observable forms and changes in the living world, and then, and not less seriously, with the process and apparatus of observation, together with the meaning of such terms as idea, thought, judgment, and mind. This course is recommended notwithstanding the difficulties which surround the investigation of mental occurrences. Limits to the understanding are encountered, whatever be the direction taken.

One Part or Character of our own Life is engaged in observing other Characters of Life

In conclusion, we may say that by the exercise of a definite part of our active existence which we call observation, we become aware of certain changes in objects around us, and in consequence of our observation of these changes, we speak of the objects as living. It is difficult to describe all the changes which characterise life, or add further to the list, without making statements for which we are unprepared, though the further statement made in this chapter, that a distinct part of our own lives lies in the faculty itself of observing, may be readily accepted.

Whether or not we may incorporate this activity of observation with a general resolution of life into change, whether, in fact, we may make the statement that life merely consists of a segregation of certain changes, and that among them, in some of the living objects at least, there are changes of consciousness,—this remains to be seen. A part, at any rate, of creation *feels* and *knows*, in addition to moving and feeding, and it ought to be clear that the current ideas about life are the product of one part of life, namely, feeling, reacting with that part which is not feeling, however mysteriously it may be connected with it. We cannot avoid this thought, and we are forced to give prominence to it. It is as well to take the trouble to understand it at the beginning. In that constant round of mental activity, which I consider to be an essential part of my total activity, the world of objects presents itself to me, and among the world of objects I may be specially concerned with living things. I myself, the feeling, knowing subject, may wish to feel and know as much as possible about all that complex network of activity which is not feeling, but is the *objective aspect of life*. The shapes and changes of living objects as they appear to me, the subject, form the field of research and the basis of speculation.

If every one will consider this as an imperfect expression of an important proposition, and will be ready to use it until he is able to substitute for it something better, the mental process which is described as jumping at conclusions will be avoided. He will be, at all events, less inclined to consider the scheme of thought, which he has derived from impressions of animals and plants, *to be adequate for the understanding of life as he himself feels it.*

Important to understand that Natural Science chiefly describes Phenomena

As we have already seen, one of the first results of the activity of the intelligence is a rough classification, whereby *thoughts* are separated from *objects*, and the appearances of things, or phenomena, are kept quite distinct from states of consciousness. This is one of the earliest distinctions drawn. It persists in almost all stages of intelligence as an expression of the collective experience of mankind, and it must receive recognition in every satisfactory attempt to understand any class of phenomena.

Scientific research, in most of its branches, deals with appearances as felt by the subject, in other words, it deals with objects as the mind conceives them. But, as we have previously stated, the complete study of the organism as a branch of scientific research, ought to deal, so far as it can, with the organism as a possible subject, capable of receiving impressions from a world of objects. Its objective representation will not be any the less faithful. Yet this treatment either implies that Biology is not wholly a science of observation, or else that the application of the methods of observation to one part of life is not justifiable.

The Organism is both Subject and Object

In its completeness the organism is two-sided, and neither side is to be neglected. But the deficiency in our knowledge of one aspect, except so far as inferences may be drawn from our own feelings, is too obvious to need expression.

Having once perceived that every organism may be both subject and object, we infer that there is something beneath the superficial distinction drawn between things that are within, and those that are without, the living subject, though a discussion which cannot be rightly undertaken now is thereby brought to view.

The Organism and its Surroundings may be considered as inseparable in some Respects

No satisfactory result can be attained, either as regards the nature of a given manifestation of life, or as regards the relation of different manifestations to one another, without considering the organism to be in intimate connection with its environment. It is a great convenience to treat the organism and its surroundings as distinct and independent phenomena at the beginning, and then, afterwards, to show the ultimate dependence of one on the other, and the relations connecting them. We may, fortunately, consider and analyse first one side and then the other, before we need discuss the nature of the bond by which they are united. But the existence of a union must be kept in mind, even though we may exercise our intelligence in attempting to disentangle, and trace to separate sources, events which are truly inseparable in nature and single in their origin.

And we must also emphasise the fact that the relation, to be assumed in this connection as existing between the organism and its surroundiugs, opens out a most important field of investigation. It has an important bearing on all the problems of life, and there is no need to point out that the solution of those problems gains a new interest in the light which it yields.

A short Reconsideration of some common Terms. Thought is based upon Sensation

A series or a succession of *ideas* is needed for the process of thought. And ideas can be traced to their elements in *sensations*, for these are the simplest facts of experience beyond which analysis cannot penetrate.

A process is the "becoming" of a change. It is a series of subordinate events which makes up a given total of change. We may or may not be capable of observing and analysing it. An object or a system of objects may change. In that case, the initial and final states or positions differ on account of a *process*. And it is so with mental states.

A sensation may be considered with regard to (1) *quality*, (2) *intensity*, (3) *extent*, and (4) *duration*. In some cases (3) is absent, *e.g.*, in the sensation of sound. The quality, however, is that which confers upon sensation its character as an elementary constituent of thought. *Difference in quality or kind is absolute; it is one which cannot be split up into minor differences.*

The sense-organs are actively constructive,

and outer changes are converted by them to sensations which vary in number, degree, and quality. And different organisms must be affected in various ways, if we may judge by the differences in their nervous structure.[1]

But Thought implies something more than Sensation

The organism is to be considered as something more than a passive subject of stimuli. If every stimulus produced an effect in proportion to its intensity, there would be no *conscious* life. If the effects of external changes made a joint impression, the result being an accumulation in which each change is fully represented, then we should have a striking resemblance to the mutual action of two inanimate bodies. The distinction really existing is worth emphasising. In inanimate action, *effects accumulate* and are independent of one another; while effects *do not accumulate*, nor are they independent of one another, in the case of living objects.

The possibility of attending to one stimulus while others are disregarded, *i.e.* the exercise of choice, demonstrates the existence of something

---

[1] And the effect of this on nervous development adds a new mode of differentiation. Each impression received tends to modify the growth and so alter subsequent impressions. The direction of growth, therefore, when once it has received an impulse, tends to become more and more specialised.

independent of external objects, namely, the mind. This is implied by the word *attention*. Again, the organism *must* attend to certain stimuli in order to survive, and it must attend to one stimulus rather than another, when both are presented simultaneously. And different organisms have different tendencies towards the various stimuli, selecting with marked constancy *certain stimuli in preference to others*. A slight appreciation of these points will obviate any risk of knowledge or thought being confused with a series of sensations.

The difficulty in making general statements about mental processes arises from the fact that we cannot feel other sensations than our own. This is so true, that there is difficulty even in expressing the contrary. The growth of language has not yet adapted itself to such a confusion of thought. *Modes of feeling are not common to all individuals in the same sense as objects are.* And herein lies an important distinction.

The World of Objects remains the same while innumerable Individuals receive Impressions from it

A primary fact of human intelligence is the belief, that objects are invariable in their effects on such individuals as are similarly organised and belong to the same class, that there is one permanent, unchanging system of objects, yielding similar impressions to a countless multitude of units of the same kind, each a complete and self-contained intelligence. What these units have in common is the outer system of objects, which acts on them similarly and thereby provides means of communication between them. The impressions made by the system, even the thoughts which, in our own case, are based upon it, when translated into action (even words are nothing but a secondary and symbolic action) are approximately alike. *The cause is single, the effects are innumerable. Where like effects occur, there similar individuals exist.* The degree of likeness in each class depends upon the nature of the impressions. For the simple sensations which we ourselves derive from objects there is absolute likeness, or what is equivalent to such likeness. It is only when combinations of impressions occur and they become increasingly complex that differences arise.

Two Ideas are aroused by the Word "Self"

There are two distinct ideas which may be aroused when I think about my "self." I may think of my own mental habits, or the conscious course of my thoughts, that is, of the mode in which outer things affect me. But, on the other hand, the word *self* may be used to describe that which I have in common with all persons, *a singleness of existence and perception.* Different species of mental *combinations* will frequently arise, though even the more involved ones may be called up by the same event in many individuals. And although our mental *processes* are mostly alike, yet each *self* has an isolated, complete, and indivisible existence.

Perceptions and Ideas

The word *perception* is used as referring to an object which is at the time producing changes in, or impressing, the subject. To perceive an object, we must have it before us and within reach of our senses.

An *idea* denotes what is remembered or imagined. It is a change produced by previous perceptions, or by combinations of previous perceptions.

Both ideas and perceptions may be traced to simple sensations occurring together or in sequence. The difference between them arises from their mode of creation. The perception is originated directly by an outer change combining with an internal change; while an idea is an internal change, aroused directly or indirectly by external changes which are past in time.

Both perceptions and ideas are necessarily produced in part by the organism, and by the central nervous system of the organism. We cannot regard them as other than *the joint product of outer changes and a sensitive organism.*

Since they may be referred by analysis to *simple sensations* as the elements of which they are made, they may be classified and arranged in the same manner as their elements; and a connected system of knowledge may be constructed for these subjective states, by the same methods as knowledge of objective states is acquired. Yet it will not be forgotten that it is *the transforming mind*[1] which gives to a perception or idea its *singleness* or *unity*, and its value in the sum of experiences.

The perception of the *self*, as *the something in apposition to everything else*, is always a late event in personal history, as compared with an experience of the objects outside the self. That which first occupies the attention, and constitutes nearly the whole of consciousness, is a series of impressions made by external objects and changes upon the passive subject. And this form of mental activity continues throughout our life. *We never cease to be influenced by external occurrences and objects, however much we may be exercised or disciplined in introspection.*

The Province of Biology

The study of Nature is not of a lower order

---

[1] One aspect in which *mind* may be regarded, though not the only one, is that in which it appears as *a permanent effect* mainly produced by *a sum of changes which are themselves transient*, namely, our total experience, whether in the form of *sensations, ideas*, or *will*.

because it is the earliest begun ; nor even because the study of the faculties by which Nature is observed, an essential in the correct estimate of the results of our observation, necessarily begins later. *Yet if we ourselves form in any sense a part of Nature, and if Nature is to be treated as a whole, we cannot admit to a very high rank a science which refuses to consider the active intelligence, while dealing exclusively with the passive material of intelligence.*

It may be agreed that we do not deal with Nature as a whole in the sciences of Physics and Chemistry, for in them we discuss *appearances* or *phenomena* as they may be, and are, perceived by all intelligent persons alike. And it may be advisable in some cases, for example, when an exhaustive and special method of observation becomes necessary, to study living changes *entirely as phenomena*, and to limit ourselves to the manifestations of life.

Many persons will consider that Biology should be confined to the reception and arrangement of such impressions as may be made upon them by living objects. But this would be an incomplete science, for it would not include the treatment of that which is a very important part of our own life, and, as may be inferred, of all life, namely, *the capacity to receive impressions from external changes.*

And more than this, the effect of these

impressions on various organisms, and the agencies by which the effects are produced, will need to be considered, for these are questions which will be sure to arise. It is better to have some authority for our answer to them, than to make those hasty inferences from the known to the unknown, which are so often not only erroneous, but also a hindrance to further progress.

A wider view of Biology is that it co-ordinates the many facts of life as they appear to us, and also attempts to gain some knowledge of subjective states. *If we find that we can learn nothing with certainty of subjective states other than our own, the attempt will, at all events, prevent us from assuming a knowledge of them which we do not possess.*

A Summary and a Sequel. There is no real Disparity between the Organism and its Environment

We trust, then, to detect in the fragmentary conceptions hitherto expressed the germ of some simple course of thought, which will lead to a more comprehensive view of the functions or activities of the organism. No safer stepping-stone to success can there be than that from which we regard Nature as divided into two parts, the *organism* on the one side, and the *environment* on the other. Nothing could appear, on the surface, to be more unequal than such a division. Yet the slight fragment of living matter which is designated an organism, requires for its elucidation, if not the whole of material nature, yet so great a part as is, in respect to itself, a world. Indeed, the environment is the world of the organism.

There may be considerable disparity, it is true, when we regard an organism and its environment as two divisions of our own objective world, but it will not do to forget that we cannot wholly study the organism as an object. We need to understand as much as we can about it in the position of a subject. It is when we consider it as being acted on by, and receiving impressions from, its own surroundings, that we realise there is no inequality, from a biological point of view, in giving as much attention to the one side as the other.

The Contrast is necessary to the Conception of Life

That there is at least a most welcome amount of convenience in the distinction which is drawn, will undoubtedly be admitted. But there is more than convenience. If we can take our own experience as a guide in such matters, we are bound to make such a contrast. We not only *act*, we *feel*. This simple expression of an obvious fact is a necessary preliminary in commencing the study of living things. It leads to the conception, here expressed, of the organism, the active being, in an environment which it feels.

Many opportunities for speculation arise from this contrast, but they have to be neglected, a fate which often befalls many of the problems arising from the groundwork of scientific thought. Even in a superficial survey of what constitutes the environment, we shall need to use terms expressing most difficult abstractions and suggesting many troublesome problems. The ideas we already possess concerning such abstractions as *matter, space, time,* and *mind* must be pressed into service. And these are but the elements, imperfect and vague though they be, in the composition of our ideas of life.

The Link between Organism and Environment

We approach now to that conception of *sensation* whereby organism and environment are linked together. There is no question here of ordinary connection or association. The union is more intimate. The word sensation in its fullest meaning may be taken to cover every kind of direct effect which is produced in the organism by its surroundings : is limited to the organism : and cannot be observed from outside it. It is the experience, evidence, or information brought through the senses or by any other route from outside.

The surrounding world has been placed in contrast with the subject which feels it, and before long we enter on a discussion of the manner in which such feeling takes place. Behind this bond uniting the two we cannot go. No instance of life suggests a simpler mode of regarding this primary fact of existence; and whether the scheme of thought based upon it be complete or incomplete, *we have nothing to take the place of this mental image of an organism situated in an environment, which reacts on it through the agency of sensation.*

CHAPTER IX

THE MATERIAL BASIS OF LIFE

Introductory

THE chapter which now opens will be mainly occupied in enforcing the ideas which are contained in two propositions. *That a general sameness in the prominent changes exhibited in common by the various forms of life, such as we have already dwelt upon, corresponds with a similarity in the substance which is the medium of those changes*—is the first of these propositions, and the second is practically a corollary of the first. It may be expressed as follows: *the diversity in the material form of living objects arises from the multiple products of the activity of a common substance, and not from a fundamental difference in the kinds of living matter.* Like most statements concerning the living organism, these propositions are, at best, but a compromise with truth, and need both explanation and amplification before their exact meaning is laid bare.

The Basis of living Changes is a Material

By bringing together the main facts which make up our conception of life, we arrive at an assemblage of events or changes connected with a material basis. *Every act of life is perceived as a change occurring in matter.* And this result is inevitable, for it is another instance of the familiar axiom, that only by the observation of material bodies can change ever be directly perceived. *Growth, decay,* and *reproduction, the assimilation of food* and *the response to external changes,* events which always enter into our perception of life, come under our observation as changes occurring in the states of certain aggregates of matter. The starting-point for all observation of living or inanimate nature is the same in this respect, that it is based on perception of matter in process of change. Hence there need be no misapprehension, when allusion is made to the material basis of life. Indeed, the relation of the observer to Nature cannot be expressed in any better fashion. There is no other figure so well fitted to embody the act of perception. Yet frequent errors of judgment arise from ignorance of the symbolic nature of such terms, and life as an ultimate fact is habitually confused with the appearances which are shaped by it.

Living Changes are few in Number and form a constant Group

There is undoubtedly a danger that a certain constancy, which must always be associated with the character of living changes, may not be clearly recognised. The events themselves are few in kind, varied enough though they may be in manner and relative extent. It is not easy to gain a clear recognition, unembarrassed by details, of the important fact, that not only this and that individual living object, but *every organism in existence*, assimilates, is influenced in a special manner by outer changes, and may reproduce its kind. There is no slight difficulty in forming an estimate of all that is included under this statement, and in bringing into review the varied changes of the animal and vegetable kingdoms which are covered by its terms.

One of the most prominent ends towards which so many of our efforts have tended will be gained, if it be clearly apprehended that all living objects, however dissimilar in form and in the minor habits of life, are alike in the possession of a definite group of qualities which separates them clearly from the rest of creation. But this point of view is not reached by a consideration of each of these qualities separately. It is dependent on our perception of them as a group of coexisting characters, jointly forming the composite phenomenon of life. No one of them can be absent, and life, in the right use of the term, still be predicable. They constitute, in a manner, the simplest chain of events upon which life can hang, though it be life in its barest expression. Much more may be asserted of individual forms, but these events at least must be manifest in every case. Life cannot be simpler than what is denoted by these conditions.

Attention to be given to the Matter concerned in the Changes

From an outlook such as we have now reached, another view presents itself. Little effort is needed to transfer the attention from the organism as a whole to the matter which enters into it. But in doing so, it must not be forgotten that organism and matter are far from being equivalent terms. The organism, it is true, is material, but it is also a centre of change and activity. Matter is merely the *substance* of the organic activity.

Additional force is given to this statement by the change of view which is now proposed. In that change of view, living objects present themselves as fragments of a substance which is, in some respects at least, the same wherever it may be encountered. *Whatever form it may take, or whatever qualities it may display, we may always* assert of the living material, that it is an assimilating, sensitive, and reproducing substance.

For this reason we do not go too far when we endeavour to bring all such thoughts together in the conception of a common substance of life. The advantages of such a conception are great. The common groundwork of fact, when once perceived, will serve as a basis for new ideas. A substance exhibiting the fewest characters compatible with life becomes the starting-point from which all the varieties of life originate. It is necessary to put aside for the present all questions concerning form and structure, all problems as to the origin of divergent modes of life, while the attention is turned towards the substance in which life is apparent.

Difficulty in perceiving the Substance as common

It is in no way remarkable that there should be some difficulty in reaching this point. The diversity of form which is observable is naturally misleading. Differences in external appearance, in the colour and texture both of surface and of dissected parts, so seriously disturb the judgment and distract the attention that common characters are not spontaneously disclosed. With such wide divergences as appear among individuals, the observer may well fail to notice the element of similarity possessed by their substance. The history of the science confirms this statement.

It displays the slow growth of the perception, that the body of the organism may have some points of similarity in all cases, and that certain qualities are present on all occasions.

Degrees in the Resemblance of Material

The question of *degree* in the resemblance ascertained has not yet been answered. The answer may take more than one form. When we come to discuss the subject from the vantage ground of Chemistry, we shall be able to express ourselves in terms derived from the theory of the chemical elements; but at the present moment a wider outlook is expedient. *The chemical treatment of the question can produce nothing more than the old result, expressed though it may be in special terms.*

In the first place, the material of the same kind of organism, *i.e.* of members of a species such as may be possibly connected by descent, may be admitted to the fullest degree of relationship. This fact is readily accepted. The partial continuity of material implied by descent would necessitate its acceptance. But the material resemblance admitted to exist between individuals, alike in all respects save those which are biologically insignificant, would only appeal to the initiated as likely to disclose itself, when a highly organised vertebrate is compared with a minute protozoon or even with a plant. These extremes appear too distant for the purpose of comparison. No recognition of kinship would occur here, were it not for the persistence, even in the lowest forms, of the same prominent events of life. Indeed the application of the term *living* denotes the possession of certain qualities, and, as we have already stated, the possession of *a definite group of qualities*, which produce on the observer in all cases *the same impression*, that of life. Nor in that case can we fail to perceive, that the impression made may be conveniently claimed as due to the same substance exhibiting its specific qualities.

The qualities common to the most widely separated examples of life are, and must be, those which are common to all. And it is in respect to these qualities that the substance of the organism never varies. We must refer to the discoveries that arise from other scientific investigations, if we wish to learn whether the substance is invariable in other of its properties. From chemical investigations, for example, we shall ascertain if constancy of one kind is borne out and confirmed by constancy of composition.

Similarity of Substance, after all, is a Mode of stating Similarity of Behaviour

But we must admit, as the logical result of our discussion so far, that variety in the manifestations of life is more justly referred to that *which is external to the substance*, and, if our judgment is correct, it is something not material. If we assert a common substance to be the vehicle of life, then an explanation of the degree, kind, or extent of life must be sought elsewhere.

It is true that similar functions imply similar conditions, and these we are constrained to consider as similar states of the same substance. Yet in saying this we merely reassert the same fact in different words. But we shall have gained this much—*an assurance that if the substance be the same in all cases, then the varied manifestations of life must be brought into connection with something external to the organism, however truly the similar manifestations may be attributed to that which is within the organism, namely, the substance composing it.*

After all, we are only able to assert sameness of substances when they yield to us the same impressions. To speak of all living objects as being *alike in substance* is nothing but repeating that they are *alike living*. The meaning justly conveyed by the phrase—a common material basis of life—is certainly nothing more than a recognition of similar material events in all living forms. To explain the resemblance in activity, by regarding it as due to a common substance, would be to mistake the meaning of the word explanation. *The terms of the explanation are derived from another impression of the same fact.*

The Process of Generalising

In the previous pages attention has been mainly directed to certain general statements which represent the results of observation, more or less controlled it may be, but, at the same time, distinctly severed from that specialised mode of observation which is usually described as being scientific. These results are obtained by the same mental exercise as is in progress at every moment of our daily round of contact with Nature. That which is distinguished in the results is the degree of generalisation. The statements are of interest, merely in so far as they illustrate the generalising process, and the abstraction which follows from it. On every occasion, when thought is definite enough for expression, it becomes abstract in form, and the form or *abstraction* stands as the generalised record of a number of similar acts of observation. This description of the meaning of an abstract term, and of the process of generalisation by which it is gained, will serve for our present purposes, and will perhaps enable us to recognise the frequency, in fact the universality, of its occurrence in thought.

Many of the generalisations expressed in these pages have been somewhat vague. There has been necessarily an absence of clearness in the descriptions which have been made. But it must not be supposed that a thought, which is expressed merely in outline and without distinctness, is consequently an inferior kind of knowledge. The larger the generalisation, or, in other words, the greater the variety of facts which are included under the general term, the greater the difficulty in giving a clear and precise value to that term.

This difficulty is rarely absent in attempting to bring our thoughts about life into connection with one another. But a gradual progress from indistinctness towards clearness may be looked for as our knowledge widens, inasmuch as we may begin then to turn our attention from *the sum of the manifestations of life* to the special phenomena which make up that sum when taken together. But as no detail can be rightly understood unless some image of a *whole* be already in existence, it is expedient to deal first, as we have done, with the outline presented by that whole, dimly though it be discerned.

Observation to be controlled and defined in Scope

In acquiring systematic knowledge of any kind the attention needs control and direction. The various items of knowledge can only be gained by a concentration of the observing intelligence upon each in turn. While one change or one quality is under investigation, no interference or distraction from others should occur; and this condition may be brought about either by a deliberate exclusion of interfering elements or a passive neglect of them.

This statement is no less true of biological progress than of other systems of research. Each aspect of living changes should be studied hereafter by itself. And it must not be supposed that any phenomenon, whether a material object or a change, will submit to the observer its aspects and qualities distinct from each other, or present them in a definite sequence. We have been furnished already with a certain number of conceptions, indistinct though they be, but to render our grasp of the subject more secure, a more active exertion of the intelligence is demanded. It must disentangle, or *abstract*, for itself, one by one, the connected facts which enter into life. *In a word, the knowledge hitherto gained can best be extended and rendered more definite, by the exercise of a special mode of intellectual progress which is described as abstraction.*

The Limitations of Chemical Methods

An attempt will now be made to deal with the material basis of life in its chemical aspect. In order to extend our grasp of the conditions of life, we shall follow the mode of investigation and inference which is employed in learning the composition of various kinds of matter, and in ascertaining the behaviour of different kinds of matter with one another. There is nothing in the nature of Chemistry to restrict its application to inanimate matter; it is concerned with all events which arise either from *decomposition* and *resolution,* or from *combination* and *rearrangement* of material. To all appearance there is nothing in the nature of living matter which removes it from the operation of laws observed by the rest of the material world. There are certain prominent changes which are observable in the substances associated with life, resembling in all their features familiar chemical changes. Indeed, they belong to the same order of occurrence. They may be observed by the same methods, and be described in the same terms as are customary in Chemistry.

Several of the facts of life may be described, then, in chemical terms, though no insight into the nature of the *connection* between chemical facts and organic facts is necessarily gained thereby. Though a wider view may be gained, it must not be forgotten that the adoption of a chemical phraseology in describing them *adds nothing in the way of explanation.* It will be seen that the final result of the operation is to bring certain of the changes connected with life into their proper association with similar changes outside life, by means of employing the same nomenclature.

K

The Application of Chemical Nomenclature

One of the prominent constituents of the matter connected with the process of life is *Carbon*. It forms part of the substance of every organism, and enters necessarily into the composition of its food. It is one of the several elementary substances which are incessantly entering into, and being rejected from, the centres of living change. In combination with *Oxygen*, carbon forms a gas, *carbon dioxide*, which is everywhere present in the air; but the greater part is fixed in the carbonates which make up a considerable portion of the solid crust of the earth. Oxygen itself forms eight-ninths of the total mass of water, and 23 per cent of the atmosphere. Besides this, it constitutes, as far as can be estimated, about one-half the crust of the earth.

Hydrogen is found mainly as a constituent of water, and, in small quantities, in combination with nitrogen as ammonia.

Nitrogen forms nearly four-fifths of the atmosphere. It is also present in ammonia and in nitrates and nitrites.

Sulphur and *Phosphorus* occur respectively in combination with oxygen and other elements as sulphates and phosphates. They are widely, though sparsely, distributed.

These four elementary substances, together with

Oxygen, are always present in living matter, and jointly compose nearly the whole of its bulk.

Chlorine, Iron, Sodium, Potassium, Calcium, Magnesium, and one or two other elements, form part, though in very minute proportions, of organic matter. They are found in various states of combination almost everywhere on the surface of the earth. Some of these compounds are directly absorbed in very small quantities by plants, and assimilated without undergoing much change.

The complementary Relation of Plants and Animals to assimilable Material

The carbon dioxide which is present in the atmosphere is assimilated by *plants*. A certain quantity of this gas enters the plant, and only a fraction of this quantity can be discovered to leave it. There occurs, speaking broadly, during the life of a plant a constant transfer of carbon dioxide to the material of the plant, and a corresponding diminution of the gas in its environment. And the process is a general one for all vegetation.

But while the withdrawal of carbon dioxide from the atmosphere is going on, oxygen is being constantly added to it. The diminution in the quantity of the carbonaceous gas is coincident with an increase of the quantity of oxygen in the atmosphere.

The part played by animals in their material relation to the environment is of a converse character. The animal, it is true, incorporates solid carbonaceous material; but it also yields to the atmosphere a constant supply of carbon dioxide, and, as constantly, withdraws free oxygen from it.

Confining our attention to the elements, carbon and oxygen, and to the compound substance, carbon dioxide, we arrive at the conception of *a process of circulation* in which the same matter may pass from a vegetable to an animal residence, and, to continue the cycle, from animal to vegetable. It is a process, however, which cannot be followed positively in any individual case. Yet the absence of any other known causes contributing to fresh supplies of either oxygen or carbon dioxide, and, indeed, the knowledge that there are other events occurring in nature which tend, on the whole, to diminish their quantity, confirms us in the belief that the two changes are complementary, and that they approximately balance one another. *Material required by the animal is supplied by the plant, and the animal, in its turn, supplies material needed by the plant.*

To give our conception a still more comprehensive form, it should be stated, that while the useful assimilable material, as a whole, is constant in quantity in spite of its varied transformations the individuals through which it passes are endless. *Animal succeeds to plant, and plant to animal, in alternating possession of the same material.*

Again, speaking broadly and with reference to averages, similar changes may be described in which the other elements of living matter take part. *Hydrogen* contained in water and ammonia is absorbed by plants, and when, after undergoing various changes, it has become part of their substance, it may serve, in common with other vegetable products, as food for animals. It leaves the animal after a variety of changes to be absorbed afresh by plants. *Nitrogen*, too, is an important element in the food of plants and animals, and alternates perpetually from one to the other, though the forms in which it can be assimilated by them are limited in number. The enormous quantity of free nitrogen in the atmosphere is never directly utilised by the organism, though indirectly, through the agency of bacteria or fungi, it may enter the composition of plants.[1] A relatively small quantity of it in combination, ammonia, nitrites, and nitrates, is in most cases alone available. The latter compounds are assimilated by plants in company with carbon dioxide and water, and they are jointly elaborated by them into complicated

[1] It is not easy to admit this complicated symbiotic occurrence within the circle of our discussion at the present stage.

substances, mainly *proteids*, which are essential constituents of animal food.

As a final result of the changes taking place with regard to *nitrogen* during animal life, it is restored to inorganic nature in the form of ammonia, or of the nitrates or nitrites which result from the oxidation of ammonia. *Hydrogen* finally quits the region of life in the substance of water; and the *carbon* contained in the proteids furnished by plants, to which can always be traced, directly or indirectly, the continued supply of this necessary article of food, is given up again in carbon dioxide.

A similar process of circulation is observable with regard to all the elementary substances forming the basis of the organism. *The same material is used over and over again in the sequence of life, the locality in which it is used—the centre and circumstance of transformation—constantly varying.*

The Names—Carbon, Nitrogen, Oxygen, etc.—are given to certain constant Phenomena

The statements contained in the foregoing section may be used in illustration of the manner in which delusive explanations arise. A class of descriptive statement, based upon one set of researches, is applied in such explanations to a class of inquiry, in the prosecution of which a totally distinct method may be imperative. And the application of chemical terms to familiar events, with which they have not been previously associated, is frequently regarded as an explanation.

But dissimilar as biological and chemical classes of statement may appear to be, they both have the same origin. They both arise from, and express, the same operations of the intelligence. In order to grasp this, it is enough to take an example. The matter which appears to be related in a special way to that which is outside it, is described as living. That which is needed in burning, or is found in water, is called oxygen. In each case a *constant phenomenon* is denoted by a definite name. It may be true, indeed, that the abstraction which precedes chemical description *is more completely analytic*, just as the phenomenon is more of a detail than the biological occurrence. In fact, the constant appearances with which certain of the names so well known to Chemistry are associated, clearly form components in the wider events of Biology. The latter, then, have been to this extent analysed, but not necessarily explained. If they could be resolved completely, and shown to contain nothing but chemical components, there would be reason in claiming the process as an explanation.

To repeat this train of thought, we may say there are certain fixed characters in the impressions made upon us by matter. The characters may be those which are designated by the names given above, characters which are elementary and constituent in kind. They may be composite, such as we recognise in chemical aggregates of matter, and again still more composite, as is the case with the impressions made upon us by *living matter*. This distinction, or classification, is one which needs to be drawn before we enter upon an investigation of the relation of the chemical events to the total activity of the organism— how far the one group of changes assists the other, or whether they merely exist side by side.

The wider generalisations of Chemistry itself, it may be here pointed out, are indeed often disguised by the unfamiliar and technical language in which they are clothed. The terms and descriptions have gained their usage in a limited and purely chemical application, and it is forgotten, when they are extended to other facts, and especially to those of life, that they gather no additional force thereby, and bear no meaning which did not belong to them before.

CHAPTER X

THE ORGANISM AS A CHEMICAL AGGREGATE

Chemical Changes described

THE customary description of a chemical change records the events attending a certain association of matter. And it records them through a system of terms, which has been shown by experience to be fitting and to encourage further discovery. But the course of chemical narrative is apt to foster a belief that the substances taking part in a change afford an explanation of the change itself. If there were less confusion with regard to the meaning of the word "explanation," this would not occur. The act of *oxidation*, to furnish an illustration, is not in any sense *explained* by the proximity of the substance oxygen, nor is it enough to describe the event as due to a property of oxygen. The change called *oxidation*, which takes place when oxygen is in contact with certain other substances is, in a very definite sense, *oxygen itself*. This noteworthy event, occurring under definable conditions as to temperature and contact with other bodies, is the basis of our knowledge of that substance, and is indeed the chief item of that part of our consciousness which we designate oxygen. Hence we are certainly justified in thinking of the substance as too intimately linked with the change to be available in explanation of it.

The so-called *elementary substances* are known by fixed and definite properties. They impress us by the same characters so long as they are in a free or isolated state. But they are capable of entering into combination or association, according to laws with the discovery of which Chemistry has been mainly concerned. In so doing, they lose most of the properties by which they are separately recognisable. It may be said that their individuality disappears under certain conditions. One prominent property, that of mass, is, however, in no way affected by chemical change. If two substances unite, their joint mass is unaltered. This is a fundamental law of chemical change, and is described as the principle of *the conservation of matter*. There is no change, of which we have knowledge, whereby matter is created or destroyed.

Yet though the more striking properties of different kinds of matter may be entirely altered by the process of chemical combination, a process of decomposition or analysis will bring about a restoration of the several forms of matter as they previously existed. That is, the process may be reversed, and the properties are then resumed. *It is certain that chemical changes are alterations, not of substance itself, but of the modes in which substance affects our consciousness.* If a substance can pass through a variety of such changes, and emerge from them in its original state, we gain little by transferring our attention from the change to the substance. It is quite clear that our comprehension of the change is not rendered more exact thereby, much as it may be widened.

These statements hold true, not only of changes in which the simplest forms of matter take part, but also of those in which more complex substances are the ingredients. It may be said with equal truth, that any kind of matter, whatever its degree of complexity, may lose for a time most of its properties as a result of association with other substances. Likewise, it may be affirmed that in all cases the old properties reappear in their entirety, the moment the original conditions are restored. There are, indeed, to be observed various degrees of complexity in the products of chemical combination, and also corresponding stages to be attained in the inverse process of decomposition.

If we proceed to apply this mode of description to well-known changes in the material of the living organism, we do not advance very far.

The knowledge is still of the same character as is expressed in the statement, that the properties of food disappear while those of living matter appear. And little real progress is made by using a chemical terminology for the intermediate stages of this occurrence. For each minor detail *is, in its own degree, a repetition of the whole operation, and resembles it fully in character.*

There is nothing truly known beyond change of property or condition. Our ignorance is still unrelieved, notwithstanding our talk of the substances as combining in progressive degrees of complexity (*constructive metabolism*), and then, on reaching a limit, becoming progressively resolved into more simple aggregates (*destructive metabolism*). And so long as nothing but a transcription from ordinary language into the chemical tongue is achieved, the search for "causes" or for explanation will be unsuccessful.

The Organism described as an Aggregate of certain definite Kinds of Matter

The simplest view of the organism from the objective side shows an aggregate of certain definite kinds of matter. The components of the aggregates are easily recognised, and may be resolved into their isolated state. But the *act of resolution*, no less than the *process of elaboration*, is perceptible only in its *attendant phenomena*.

And these phenomena are, for the most part, such as would be apparent in a change of the same kind carried on in the laboratory.

But there is at least *one important condition which distinguishes the progress of an experiment from the practice of the organism.* The experiment in the laboratory is not a spontaneous act. It takes place under external control. The agency of intelligence is needed, so to dispose the material and arrange the conditions that the change may come about. The organism, on the other hand, contains within itself both the changing material and the controlling agent. *The changes in question form part of its characteristic activity, and it is self-sufficient in its control of them.* But another distinction stands out still more prominently, that the outcome of chemical changes in the organism belongs to a new order of events. It creates a fact of life, not of Chemistry.

The multiplicity of these changes, and their classification, does not form a part of the scheme before us. It is sufficient for our purposes to take their resemblance to other chemical events for granted. There are, proceeding simultaneously, in every living organism many events which tend to the construction of elaborate aggregates, and likewise many events which bring about the resolution of material so built up. A metabolism which is two-sided brings about in one place a disturbance of structures that have been raised elsewhere.

Chemical Description does not cover distinctive Characters

But whether the products of these changes are simple or complex, whether ascending or descending in the scale of metabolic service to the organism, the act of giving chemical names to either changes or products must not be considered *to create knowledge that is other than chemical.* They may be described in chemical terms, such as apply with equal force to changes and material which have no connection with life; but inasmuch as the terms do so apply, they can have no specific biological value. *The province of Chemistry extends, it may be truly said, to Biology, but its principles are constituent, and not inclusive, in that region.*

When, too, we come to consider the properties of living aggregates, we find similar distinctions to be maintained. Inanimate chemical aggregates present properties which are unlike those of their components, and biological aggregates likewise manifest qualities which are distinctive and *sui generis*.[1] It is only when we take a broad survey

[1] The general sameness of chemical constitution must be accepted for protoplasm, but a great variety of physical state must be admitted. Without difference of arrangement or condition, it would be difficult to understand the varying effects produced by stains, poisons, and other agents upon the living substance.

of the activity of the organism as a whole, that the force of the distinction is fully felt. Just as the aggregate of chemical elements impresses us differently from the elements themselves, so the phenomena in the organism which may be resolved by one method of analysis into chemical processes, yield in their *accumulated effects* a result of a new order, and one which is certainly not chemical.

The methods of Chemistry may be employed to classify the events and the substances which contribute to the activity characteristic of the organism. But the activity in question is something more than a *mere accumulation of material processes. Indeed it is not so much an addition of one process to another, as the utilisation and the control of processes which constitute the fact of life.*

A Description of Matter in General Terms

To express or attempt to express any given thought often tends to make it clearer, though there may be no immediate addition of clearness or lessening of indefiniteness. Clearness more commonly comes from the frequent repetition of an impression under varying conditions. As a rule, indeed, we add to the number of ideas which are collected around a thought, whenever we use words to communicate it to others. For many words carry with them *a composite idea* or *judgment*, which has been refined and rendered definite by long usage and frequent comparison with varying impressions.

In attempting to express the thoughts which are recorded by the word *matter*, we find this statement well illustrated. There are exceptional difficulties in this case, for a definition which will satisfy the requirement of all schools of thought cannot be attempted with any hope of success. The most diverse views are taken, according to the mental habits or training of the individual thinker. To one person matter suggests itself as having *a real existence;* to another it is *unreal* and *illusory*. It is regarded by some as *a creation of the mind*, by others, strange to say, it is looked upon as *the cause of mind*. Without attempting a conciseness which is not called for by the inquiry before us, it may yet be possible to associate with the term matter such a meaning as will enable us to continue our course, while realising that our first steps are always uncertain and liable to revision.

We find it necessary now, as on previous occasions, to make use of a provisional definition, and we have to allow the word *matter* to stand as a sign of a certain working hypothesis. We must agree that a given idea, which we know to be present in our minds, shall be symbolised by that word. By so doing, the word is not precluded from subsequent use in other senses, nor is its meaning intentionally limited. Just as the epithet *living* may bear very different meanings to different intelligences, so must it be granted that the word *matter* may recall to the consciousness of different persons very varying effects.

There is, however, one clear and distinct phase of our conscious experience which remains the same, however much we may vary the description of it; and it is unalterable as a fact of experience, however dissimilar may be the interpretations put upon it. *This primary fact is not, indeed, the certain knowledge of the existence of a material world outside ourselves, but the existence of states of consciousness which appear to be, and may be, produced in us by an external and material world.* The effect produced in the conscious being is that which we shall describe as matter, and not that unknown something which we try in vain to understand. We are able in this way to rid ourselves of all hypothesis as to the real nature of matter—whether we see it as it is, or perceive an image of something else, or indeed whether it is truly outside ourselves or not—for we confine our attention to a reality, *a state of consciousness.*

Our knowledge of matter is both a sequel and a record of states of consciousness, but it discovers no ground for saying *that matter causes consciousness,* or *that consciousness causes matter.* The basis of a connection may possibly be laid bare to inquiry hereafter, but all that can be said now is found in the proposition, that as long as life exists they are *indissolubly connected*. The link is unbroken during phenomenal life, and the biologist has nothing further to add to this statement.

Sensations are real at all Events

The sensations of *hardness, colour, cohesion*, and so on, are realities for us; and such sensations, whether separate or combined together, make up our experience of an outer world. It is permitted to be doubtful about the reality of the material world, *whereas our sensations are most clearly beyond doubt*. Nevertheless, it is important to remember that the attribute *material* is used quite legitimately in this connection, for it serves as a symbol of an absolute fact of our existence, that variation of feeling which we refer from necessity to an external world. For every case of conscious life, this is the most convincing formula.

It is methods of perception that fix the forms of scientific research; and we must leave to others the discussion of matter as a reality, or a cause of perception, giving recognition again to the axiom, *that such problems as the existence of matter in the absolute do not come within the province of Natural Science.*

It may be gathered from these statements, however, that the explanation of external nature must be sought in the character of the organism which perceives it, and, likewise, that the true significance of *the organism* must be extracted from a conception of *that whole nature of which it forms part*. The division of Nature into two parts, one living and the other lifeless, is convenient and, for many purposes, necessary. Yet neither part can be adequately conceived by itself. For the present, in default of any wider conception of the whole, we are driven to regard the organism as giving, in the reality of its sensations, the most consistent reflection of an outer world, and to find in this reflection our most profitable field of inquiry.

The province of Natural Science assuredly does include such topics as those which deal with *sameness in the kind or the quantity of matter*. Here we return to something about which we may be certain or uncertain. An inquiry into the similarity or dissimilarity of two portions of matter is a legitimate one for the scientific student, for, in truth, he is concerned with sameness or difference of perception.

It will be no slight gain that will accrue from mastering this fundamental truth. The issue of this research will then, at any rate, be free from the obscurity that surrounds such problems. We may leave on one side all discussion of things in themselves, and concentrate our attention on something about which certainty can be asserted. And our perception of Nature is something undoubtedly real and finite. Inadequate it may be, but unreal it cannot be.

Some Knowledge of material Objects is exact, namely, Perception of Sameness

If there is one phase of our knowledge of the material world in which we may make certain and precise statements with more security than another, it is that which deals with questions of similarity. In treating of such matters we adopt one of the chief methods of scientific discovery. The perception of Nature as a whole, or of any part of Nature, may differ in different persons. There is no exact standard for comparison between them. Similarity of such perception, it is true, may be inferred from general sameness of conditions—education and character, for example; but further we cannot go. Still, on the other hand, isolated acts of perception on the part of the same person may be compared with one another within certain limitations. *If I can be sure that two states of my consciousness are alike, I can make that similarity a medium of communication with another being.*

It is far from necessary that similarity in two acts of perception on the part of one person should indicate similarity on the part of another person, even when every circumstance is unchanged; but similarity of perception is something *about which propositions can be made*, and for that reason thought can be communicated from one to another. It is not to be denied that the proposition—I feel pain—is one of a large class for which some degree of equivalence may be inferred in all cases. But the equivalence is not of the same order as that which the proposition—two and two make four—bears to different individuals. This perception of sameness is the basis of all intelligent observation. Instances are familiar. The reliance upon the coincidence of two marks in physical measurement, the recurrence of the same properties—*density, colour, degree of stability*, and so forth—as a sign of the same kind of matter, all point in this direction.

All Knowledge of Nature based upon the Perception of Similarity of Quantity or Quality

We may proceed, then, to the general statement that coincidence in *time, space,* or *quality*—all, in fact, that it is the function of science to investigate—has only become a subject of common investigation and common knowledge for mankind, because sameness of perception can be estimated by all and communicated from one to another.

Our acquaintance with the principles of Physics should have impressed us with the first great lesson to be learnt from that branch of science, namely, that all physical changes need to be measured as quantities, and that they can only be measured by means of a standard of the same kind of change. It may easily be granted that the measurement of change by reference to another change of the same kind implies no great insight into the mechanism of those events which are being investigated. Nor, indeed, do we add much to our knowledge, though we may gain greatly in order and arrangement, by a system of measurement which reduces all physical occurrences to changes in determinable magnitude of *length, mass,* and *time.*

Our acquaintance, too, with the principles of Chemistry should have taught us, that in considering chemical events, attention must be given not only to the physical changes, but to the *quality* or *kind* of matter associated with those changes. In physical observation, then, we are chiefly concerned with *quantities*, and quantities in the main of *length, mass,* and *time;* while chemical events embrace changes of qualities as well as of quantity.

In each of these branches of Natural Science, including, as they do, the whole of our knowledge of inert nature, the one prominent instrument of progress is that to which allusion has been already made, viz. the certainty of our knowledge that two separate perceptions, whether of quantity or quality, are alike. The whole course of attainment in these sciences starts from this certainty;

and the dissemination of knowledge is made possible only so far as the recognition of sameness is communicable from one intelligence to another. In applying these considerations to the subject now engaging our attention, we are forced to admit the same principle exists.

Test of Sameness applicable to living as well as to inert Matter

Our knowledge of the sameness of the matter presenting to us the qualities of life is as exact as that which is based upon any act of observation. We know it in the same sense as we know any simple chemical fact, *by the similarity of value which all portions of living matter yield to perception*. It would greatly simplify the arrangement of our ideas concerning the manifestations of life, if we could bring chemical methods by themselves to demonstrate the kind of matter connected with these manifestations. It may, indeed, be assumed, on the ground of chemical authority alone, namely, the evidence of analysis, that this matter is constantly the same in kind. Similar experiments, that is, *the same directed inquiries*, produce similar results. Wherever matter, which has at any time been alive, is subjected to chemical observation, it has been found to possess as its chief ingredients those kinds of matter which are denoted by the names *carbon, oxygen, hydrogen, nitrogen,* and *sulphur.* And this in itself would be an indication of its constancy.

The material which is associated with life, and the changes in that material which are recognised as vital changes, may be investigated, and for some purposes are necessarily investigated, according to physical and chemical methods. The substance and its changes, under such observation as can be given, may then enter into the list of chemical and physical facts. Yet protoplasm, in the light of Chemistry, it is important to notice, is *inert*, or rather, Chemistry deals only with the dead material of protoplasm and with the extinct products of life. When analysis begins, life is ended; and meanwhile the physical accompaniments of life may be looked on as distinct in some respects from other physical occurrences in Nature. *The treatment of protoplasm, then, which may be convenient for the detection of detail, gives little assistance, except indirectly, towards a comprehension of life.* Nevertheless, such is our mental constitution that we are compelled to use *the same mode of reasoning* in Biology as in other sciences.

A Primary Restriction of all Observation

Any attempt to gain a comprehensive view of living changes is always rendered difficult by reason of the limits to which all observation is confined, whether it be of living or inanimate things. In the exercise of observation alone, nothing but the objective side of life is viewed, and to that side we are confined most strictly and most obviously when we pursue the methods of Chemistry and Physics. The vital changes observable in matter cannot, of necessity, include changes of consciousness. It is unnecessary to dwell upon the impossibility of observing a feeling or thought by the ordinary route of sensation. To speak of the observation of consciousness would be to use a meaningless combination of words. Matter, and certain changes in matter, we can perceive and observe, and with these changes a form of consciousness may be associated in our imagination. The *effects* of consciousness may certainly be perceived, but nothing beyond. The presence of life is inferred from objective phenomena, and we are incompetent to apply other direct means of identification. *This restriction of view should be enough in itself to guard us against giving undue prominence, in our thoughts about the organism, to material appearances which must of necessity form the less significant part of life.*

Protoplasm something more than a Chemical Compound

The activity of protoplasm, rather than its constitution, is, after all, that which chiefly concerns the biologist. The constitution may be a subject for chemical inquiry, and researches into its changes may be instituted on physical lines, with the result that material details are learnt and the various stages of the processes accompanying life are elucidated. But though forms of matter once thought to be characteristic of the organism may be manufactured in the laboratory, and so brought into harmony with the products of operations dissociated from life, and though the details of an apparent mechanism, with every material, thermal, and electric adjunct, may be explored, the organism remains, in its purely biological import, unexplained and inimitable. Its vital qualities belong to a system of subtler events, and they exact for their adequate description a fuller scheme of thought than that which is founded on chemical and physical procedure.

Ideas based for the greater part upon our own experience of life, preconceptions though they may be, must be added to the conception of a material protoplasm before it can become truly representative. Or, to express the same thought in other words, it may be said, that after the methods borrowed from the analysis of inanimate Nature are exhausted, there remains a residuum of fact which is practically untouched by them. *This unsolved problem is the control and adaptation of every element of activity by the organism to its own end.*

The constant Features of living Matter do not preclude Variety of Form

In giving the common name of protoplasm to any portion of matter which manifests the activity of life, we are performing an act of generalisation which is conspicuous by the extent of its range. The word *life* is used to describe a certain collection of qualities, and since these qualities are apparent on all sides, the adoption of a common name for the material presenting them must be permitted on the ground both of necessity and utility. Provided our observations are correct, and the process of life has similar elements in every instance, then the use of the same word, *protoplasm*, to describe the substance of life is justifiable. *Yet, inasmuch as the whole universe of living things is included in the scope of that idea, it fails to assist us with information about the modes and forms of life.*

It remains, therefore, to be seen how far this conception of a common substance can be reconciled with the diversity of form which it covers. For those forms which may be connected by descent, members of the same class or species, for example, there is no difficulty. The fact of descent implies by itself some partial material continuity; while the similarity of habit, and especially of *all material relations*, points in the same direction. But there is, on the other hand, such wide divergence in the appearance and habits of some forms, notably when we begin to compare plants with animals, that it is difficult to believe they have anything in common.

In many cases the difficulty arises from confusing objects which are really distinct. The substance which is under observation may be erroneously considered as alive, whereas it may be only accidentally associated with living matter. In the vegetable kingdom it is commonly found, that most of the products of changes attending assimilation do not pass away from the organism, as is customary with animals. Yet it must not be supposed that their retention is prejudicial. On the contrary, the woody material of trees, just as the bony skeleton of animals, plays an important part in their welfare. Taking the subject in its broadest aspect, we cannot help coming to the conclusion that much of the observed variation of form is misleading. *Its biological significance is not really great, either with regard to distinct organisms or to parts of the same organism, though it may serve as the subject-matter of classificatory science.*

The Difficulty in localising Attributes

In considering the material of life, and in coming to the conclusion that its constitution and its activity is, in general terms, the same everywhere, a certain amount of difficulty may be encountered in attempting to distinguish the qualities of the organism as a whole from those of its parts. The conception of a common substance would naturally encourage the belief, that the properties exhibited by the whole organism extend unchanged to all its parts. The tendency will certainly be lessened, if it be recognised that the word *part* may be used in various senses.

At the present moment we are face to face with two distinct ideas, each denoted by the same term. The behaviour of the organism as a whole may, or may not, resemble that of any given portion. Every portion of living matter, whether existing by itself or entering into the composition of a larger whole, must possess that *minimum of qualities* which has been described as coincident with life. And if this be true, there must be an equable distribution of some functions throughout all matter that is actively alive. But, on the other hand, it would be far from reasonable to imagine that certain biological characters, those, in fact, which are doubly and even trebly composite in kind, extend to parts which are constructive in their nature, or to processes which are subsidiary and component. *The biological whole has been shown to be greater than its chemical or physical parts.* But more than this can be said. The union of chemical and physical characters and events yields items which are *themselves elementary for biological purposes.* A further combination and adjustment must take place before we reach conditions favourable to life.

The organic whole is not even a collection of organic parts. The meaning of the word *part* in Biology does not carry with it any idea of locality. It is not, in other words, a relation of space or magnitude, but rather one of constitution or aggregation. The thought of partition brings into prominence a completeness which is unrevealed by the material presentation. *With this understanding it needs no special discernment to perceive, that each part, and each organ or collection of parts, may have not only a subordinate share in, but also a partnership on equal terms with, the activity of the organism as a whole.* It may also be affirmed that there is no necessary connection between the operation of life and the disposition of its substance.

CHAPTER XI

THE ORGANISM AS A CENTRE FOR THE TRANSFORMATION OF ENERGY

Physical Methods and Results may be applied on certain Conditions

THE evidence which satisfies the chemist as to the constitution of the living material, is profitably consulted by the biologist; but as before said, that evidence cannot by virtue of its character, nor by reason of the means used in collecting it, furnish a complete explanation of biological behaviour. It is valid within its own sphere alone. With this understanding, the methods of Chemistry may be followed for the purpose of learning the numerous details in the varieties and combinations of organic matter. Their number and importance is amply sufficient to form, as they do, a special branch of research.[1]

But the most exact analysis of the chemical changes in question cannot be expected to throw

[1] Physiological Chemistry.

light upon their initiation or their end, much as it may inform us with regard to their course. A series of chemical events, such as are in progress during the life of an animal or plant, has no parallel, except in an elaborate course of experiments originated and controlled by an external agency. There is no circumstance, *excepting that of association with living matter*, nor indeed any known group of circumstances, which is capable of giving rise to chemical changes of such involved complexity as we encounter in the organism.

At the same time it is known, that purely chemical changes cannot be accurately described, nor indeed can we be directly cognisant of them, without reference to the physical events which accompany and reveal them. It is necessary, therefore, to seek for biological information at physical as well as chemical sources. And as far as physical events in the organism are in question, there may be noticed, with regard to them too, a singular complexity which marks their separation distinctively and prominently from every other group of physical occurrences. But this is not their only distinction. *The remarkable spontaneity, which is a feature of the chemical changes in the organism, necessarily appears again in the well-defined and balanced group of physical changes.*

The last stage of analysis with regard to matter is reached, when it has been resolved into its chemical elements; and the most abstract form in which physical changes can be presented is one in which they are expressed as alterations of definite magnitude in *energy*, either kinetic or potential. And these, we know from our physical studies, are resolvable into magnitudes which can be expressed in terms of length, mass, and time.

The surprise with which general opinion received the proposition, when first made, that living objects were formed of material similar in its constituents to the rest of the world, has had its counterpart in the slowness and difficulty with which currency was gained for the belief, that the changes occurring in these objects resembled in most of their particulars the general course of natural events. *The energy of the organism*, after being regarded as a phenomenon standing quite by itself, has been finally brought into comparison with other manifestations of energy, in spite of the characteristic differences which remain to be described.

Although there is in this accession of knowledge an apparent revolution of thought, nothing in reality beyond an extension of the field of observation has been needed to bring it about. The methods in which various physical and chemical changes are measured by standards common to all, are now applied as far as possible to certain component processes of biological change, *and the doctrine of energy is thereby discovered to have*

M

a wider application than was first conceived for it.

The Classes of Change to which Physical Considerations are to be applied

The most comprehensive image of life is one of change, and of change in its most conspicuous forms. So ingrained is this conception that unusual motions in inorganic bodies are frequently referred to living agency. A wider acquaintance with Nature shows no reason for altering this attitude. But since it would be difficult or impossible to arrive at any conception of the material world from which change is absent, and because, in speaking of all the facts of life as changes, the process of generalisation is pushed to its extreme limit, the contrary path of analysis must be followed to gain more exclusive information.

To arrive at this end our attention must be concentrated upon the changes themselves, and questions of *matter* must be, for the time, put on one side, without forgetting, however, that the observation of any event, if it is to be complete, involves some measurement of the quantity of matter engaged in it.

The first result of superficial observation may be recorded in the statement, that in life *many kinds of change are in progress simultaneously, and that each kind is made up of many successive phases.* The processes are numerous and complex. But, on the other hand, we perceive with equal certainty that the number and complexity of changes in the living organism have *an observable limitation.* Just as the matter associated with life, either as food or as the medium of activity, has a special and determinate character, so have the changes which are possible in that matter a restricted range. *They are either such as accompany the maintenance of the substance of the organism, or alimentation; or they are concerned in reproducing the form of the organism, that is, in multiplying the centres of growth; or, lastly, they have to do with the control of such activity, and with the regulation of the whole sum of activities in relation to environment.*

The latter function, which is correlating in its effects, is the most difficult to comprehend, and still the most distinctive in character. In a large portion of the living world it is rendered still more problematic by the absence of any visible organisation with which it can be associated. Under some kind of control, whether observable or not, each change, and each detail of change, contributes its adjusted share, and no more, towards that total sum of actions which we term life. *There is manifest, between all occurrences in the organism, a bond of mutual dependence, which gives rise to the conception of a system of activities combining successfully to attain a single end, which can only be described as the maintenance of life.*

The Conception of Energy

That there are two classes of things, *forms of Matter*, and *forms of Energy*, existing in Time and Space, is the most generalised statement which can be derived from the study of Physics, a science dealing exclusively with the physical universe as it is manifested to us directly or indirectly through the senses.

The student's researches in Physics have for their end the measurement and classification of all phenomena in terms of matter and energy as conditioned by space and time. Yet nearly all that he knows of matter is gained, it may be observed, from changes affecting his sense-organs, and these changes themselves are ultimately resolvable into a special mode of transference of energy from one portion of matter to another. *The first stages in the event of sensation are part of the physical world in which sensation is exercised.*

Energy is never observed except as a state of matter. Matter is never destroyed or lost, and its parts can be traced through their numerous transformations, unchanged in quantity and only temporarily modified in kind. On the other hand, there is no such permanence in the parts of energy. The total quantity of energy, as that of matter, does not change, but *no identification of its parts is possible*. Effects are apparent while it is undergoing a change in form or position, but we know nothing of it, except through, and at the moment of, its transformations.

The transference of energy from one body to another is termed *the performance of work*, and it is taking place incessantly around us. The investigation of its various forms gives rise to the several branches of Physics. But they are all alike in their subordination to the theory of a *Conservation of Energy*, which gives consistent results in all directions. It is a final generalisation or principle to which all known physical changes can be referred. More than this, it has brought new facts and relations into view, and has simplified our comprehension of all physical transformations.

Now it may be quite true that the foundations of scientific thought are vague and elusive. *Matter, Energy, Space*, and *Time* defeat our attempts to confine them permanently within the limits of a verbal definition. Yet if we agree that any one of these words shall stand for a definite idea, it is not difficult to construct equally definite conceptions for any of the other words. They form a quadrilateral of thought which must survive or be destroyed together, for each depends upon the others.

An agreement of this kind has been made

with the most satisfactory result, and it is by results that the value of scientific methods must be judged. Upon the physical conception of Matter, Energy, Time, and Space there has been raised an organised body of knowledge, every part of which is connected with every other part. No disturbance in the proportions or relations of parts of this knowledge to each other can be effected by an alteration in the definition of these fundamental ideas. *All scientific knowledge assumes ultimately the form of an equation, and an equation is not affected when both its sides are treated alike.*

Biology may, then, safely accept the facts of Chemistry and Physics. *A change in their postulates and axioms can be readily extended, if need be, to their applications in Biology.* Still if it is a science in its own right, it is by reason of facts, or relations between facts, which are neither chemical nor physical.

Energy of Animals derived, directly or indirectly, from Plants

It is usual to describe the exchanges and transformations of energy with regard to the organism in the following manner. The green plant, by aid of the energy of sunlight, forms complex material from carbon dioxide and water. Energy is absorbed in this constructed material, and resides there in its so-called potential form. This and other material constructed by the plant is used as nutriment by the animal. By the intervention of oxygen the complexity of this material is destroyed within the animal, and the simpler substances, carbon dioxide and water, are reproduced. In this breaking-down, consumption, or destructive metabolism, whatever we may term it, energy previously potential is wholly or in part transformed into the kinetic form. The animal performs work and becomes warm, or grows in bulk. A definite accession of energy is perceptible in these changes, and there is no source, beyond the store of energy in the plant, to which it can be traced.

But it must not be supposed that this statement approaches completeness. It describes those changes in respect to which the animal and plant *supplement one another*, and maintain an *equilibrium* with regard both to *material* and *energy*. Nor must it be imagined that the change from potential to kinetic energy implies any real discontinuity of action. This assertion is supported by physical considerations as well as the fact that truly living matter is much the same in each class of organism. At the real centre of events, differences depend more on relative quantities than on qualities. Oxygen is consumed and carbon dioxide is excreted to some extent by both.

All the processes of organic life bear the same class character, though they may vary in detail, and the difference may be so great as to find its expression in the divergence of the plant from the animal. Construction and destruction of material are essential forms of activity in every organism. In the plant, construction preponderates; in the animal, destruction is more prominent and in keeping with a larger performance of work.

On the whole, energy is observed to be transferred from the environment to the plant, while there is found to be a transfer to the environment from the animal, provided all the transactions taking place have been correctly observed and balanced. It would be easy to imagine the vegetable world proceeding to a maximum development, and only ceasing when the supply of carbon dioxide is exhausted; but it would be difficult to realise the animal world, as it is now constituted, continuing, or even beginning, to exist, apart from the food which plants supply.

The Organism not an ordinary conservative System

Every living organism may be regarded as a *centre* at which energy is being constantly transformed. It is by the nature of this transformation that we recognise it as a living organism. But when we come to investigate the circumstances or conditions of the event itself, we find that they are, in some points, unlike the antecedents of all other exchanges of energy.

The continuous operation of these transformations in the region of the organism is distinctive. In all exchanges of energy between inanimate bodies there is a speedy attainment of equilibrium, *whereas the organism, so long as it lives, is incessantly disturbing the equilibrium which should otherwise arise between itself and its environment.* In other words, living organisms are not ordinary conservative systems, and the extent to which they diverge from the principle of the conservation of energy is another indication to us, that in the organism we come in touch with phenomena which are not yet, at all events, reduced to physical laws.

The special phenomena of life, to which the principle of the conservation of energy has been applied, are those of nutrition and, concurrently, those of objective activity. The sum of the work performed by an animal or plant on its environment is derived from the potential energy of food or the kinetic energy of sunlight. There is apparently no animal activity (including in that term the maintenance of a given temperature in the body) without a corresponding disappearance of chemical potential energy, which may also be described as the consumption of nutriment, or as an oxidation and simplification of that matter. Nor is there any apparent source for the needful accumulation of potential energy in ordinary vegetable matter beyond the energy derived from the sun. Experimental evidence of this statement is not easily obtained, but all the evidence which exists tends to prove its truth. *We may therefore consider it as almost certain that all observable activity in plants, and in animals too, since they are dependent upon plants for their energy, is derived from sunlight.*

Resemblance of Organic Activity to that of other Systems

There is little doubt that the laws which express the relations observed to exist between mutual changes in the inanimate world do partially extend to the region of change which has just been described. Chief among those laws is the conservation of energy, and great importance attaches to it because of its representative character. All physical systems exhibit in the mutual action of their parts a strict adhesion to its terms. The biological issue before us is whether or not the organism is related to its food and to the changes in its environment, in the same way as the parts of other systems are related with regard to exchanges of energy. If it be, then, as far as objective occurrences are concerned, the living state of matter is comparable with thermal or electric states, all of them being energised or energetic conditions of matter. And there can be little objection to the description here suggested. Living matter is most patently energetic, or, as we may say, the organism displays a special kind of energy.

A description of the various changes which are subsidiary to the derivation of energy from nutriment would necessitate a long catalogue of events. But however numerous they may be, and lengthy as the list of variant modes may be, one of two ends is reached, according as the plant or the animal organism is under observation. *Either potential energy is stored up in the form of complex but unstable material, or work is done at the expense of such material previously constructed.* One or other of these results is obtained. One or other may predominate, or there may be, as occurs in some cases, a balance between them.

No complication of organisation should be allowed to disguise the beginning or the end of this process. The structures employed in it may be highly elaborate. A complex alimentary system, with various glands and a vascular system, may take part in it. Leaves with their chlorophyll and respiratory cavities, together with the vessels of root and stem, may serve in the supply of food to plants. On the other hand, the same results may take place without any organisation visibly contributing to them. And between these two limits we have interposed the whole range of zoological and botanical variation in digestive structure.

Now, evident as it may be that a great part of organic variation corresponds, with differences merely in the process of alimentation, and not in the end which alimentation subserves, there is still another important share of variation to be ascribed to a diversity in the methods of procuring food. The habits and especially the mode of locomotion (if locomotion be a necessity) of the organism, whether terrestrial, aquatic, or aerial, are almost wholly connected with its search for food. Altogether, then, much of the diversity of form to be observed, both as regards internal structure and the more striking features of external shape, can be traced, sooner or later, *to a foundation in the special relation of the organism to its environment in respect to food.* And this relation is pre-eminently one in which the principle of conservation of energy obtains. There need be, therefore, no hesitation in saying that a large part of the activity, and that too the most prominent to observation, resembles ordinary physical occurrences in most respects.

All material changes within the organism itself, and all material transactions taking place between plants and animals, or between the organism and its food, are indistinguishable in their essentials from other chemical and physical events. *The internal balance of the organic world is, in short, that of a conservative system.*

Distinction of Organic Activity from that of other Systems

But life as a whole, and the organism in its complete round of relations, includes more than processes of alimentation and such activities as are accessory to it. Without encroaching on the field of subjective facts, distinctive as they are, in spite of being in the main vaguely known and at best inferred, we have before us *an intermediate region of automatic action and co-ordination*, which remains isolated from that which is subjective, and yet separate from the ground previously covered by observation. In the wider treatment of the organism as a system of processes which are self-maintaining and self-directed, with or without the mediation of sensation, we gain nothing by referring to the laws of energy. They entirely fail to establish a satisfactory connection between the organism and the physical changes which are external to it. The innumerable events which form the antecedents of our own sensation are but seldom related in a quantitative manner to the effects they produce. Such alterations as the organism may itself produce on its environment, by means of its store of energy, are always to be classed as performances of work, with corresponding expenditure of energy.

But the effect of the environment on the organism is not susceptible of treatment as a physical occurrence. It is a problem of pure Biology.

Nor is this all. The familiar processes of growth and development receive no aid from the science of energy. They cannot be resolved into simpler conceptions by its means. The growth of a complex organism and of a single cell are at one in being outside the range of the conditions usually obtaining in the transformation of energy. Mere increase of bulk, it is true, is commensurate in some mode or other with the matter assimilated, but the act (or capacity) of assimilation and the process of development belong to a distinct class of events.

Still more justifiable is this judgment when we come to consider those occurrences which *in their objective aspect* are called reproduction and death. In dealing with these events as a normal part of Biology, we are at once brought face to face with the need of new ideas and illustrations. Their range is too extensive and their suggestions too numerous for material or mechanical analogy to be helpful. *The individual, with its co-ordination of actions and its single relation to its world, makes demands upon other sources for an explanation of its beginning and its end.*

CHAPTER XII

CERTAIN ASPECTS OF FORM AND DEVELOPMENT

The Need of Guidance in considering Form

SOONER or later a limit is reached in the mere accumulation of facts. Yet it is always advisable to enlarge experience as far as possible, before attempting to understand the generalisations which are founded upon it. The conception of form as applied to the organism takes in our consciousness a shape so unstable, that we search instinctively for something less variable upon which to fix our attention. More and more examples of living forms may be collected, they may be dissected with increasing precision, and an adequate terminology may supply names for each object and detail, but the want of connecting links is still felt. The capacity of the mind to assimilate impressions has its limits. *Comprehension must come to the aid of experience in order to render it effective and profitable.*

The whole recorded knowledge of organic form

cannot be reviewed by an individual person without the aid of those outlines of classification, within which successive generations of observers have learnt to concentrate their descriptions. Hence it is usual for the examination of certain *representative forms* to serve as a needful preliminary to the reception and realisation of the more important principles of life. The numerous and diversified classes, which these forms represent by standing to them in the relation of *types*, are most quickly apprehended by following such a course. *For thereupon a real experience is acquired, with which an assumed experience can always be compared.*

It may be also noted, that after a certain stage in mental history has been reached, the perception of *relations between known phenomena* becomes more important than the pursuit of *fresh phenomena*. And in the stage of organic research which follows the dissection and study of such forms as are usually selected for a course of Biology, the first necessity is to grasp the scheme, of which the objects form part and with reference to which they were selected.

The schemes and principles which may be presented to the student are, after all, nothing but *abstract ideas* based upon the generalisations of previous observers, and their value lies mostly *in condensing the description of the multiform* into a single phrase or term, and that often a simple one. But the condensation and economy of description which an accurate principle brings about frequently escapes notice, so slowly and laboriously do single terms fit themselves to a multitude of phenomena, and so readily is it forgotten *that such terms follow and recount the observation of appearances rather than precede. and explain them.*

The Meaning of Degree as applied to Problems of Form and others in Biology

The question of degree is one which will frequently recur in any discussion of the variety of form displayed by the organism, and it is one for which we need to be prepared. An intelligent grasp of the fact that a difference of degree is not a difference of kind, obviates many misunderstandings, both as regards the meaning and value of form and the nature of the activities associated with it.

Whenever we come to deal with diversity, enumeration is the first instrument by which any approach to order can be gained. A simple process of counting the number of distinct variants is the first step in arrangement; and the process of naming, which usually follows the recognition of different individual objects, is a further step in the direction of order. But it is long after these steps, and after tentative schemes of classification and arrangement have come into existence, that the attention is first drawn to *the degrees of difference* between objects rather than *the number of differences*. The process of comparison follows at some distance that of enumeration.

But the process of comparison is one which depends for its exact performance upon the existence of a standard of measurement; and any comparison to be of permanent value must show a quantitative result, expressed in terms of this standard. It is not always possible, however, to confine the meaning of degree to the result of a process of this kind. The term cannot always carry with it a direct sense either of quantity or number. It may simply express the extent of divergence from, or convergence to, a fixed standard. In many instances, too, it refers to a quality rather than a quantity, and the quality may be considered to extend from zero to an indefinable maximum.

Much of the variation in the forms of life is of this character. Diversity is apparent in every direction and in every degree. Yet specific forms have their counterpart, not in a diversity of activity as might be imagined, but in a diversity of degrees of activity. For the elucidation of this point, if for no other reason, it would be important to describe the ordinary usage of the term. It is additionally important, when we realise that the conception of degree, as distinct from kind, enables us to bridge over the gap between animal and plant life. Great as the difference in their appearance may be, we cannot perceive in the activity displayed by them any difference beyond that of *degree*.

The activity of the organism is less varied than would appear to the superficial observer. In other words, there are but few classes of change exhibited by it, though the mode and circumstance of those changes may give rise to innumerable modifications of each class. Movement may range in indefinite variety from the slow and partial change of position, perceptible with difficulty in plants and some animals, to the rapid flight of birds. The ingestion of food, towards which, as a rule, locomotion is preparatory, may vary from processes in the plant which are only observable with difficulty and by indirect means, so imperceptible are their gaseous and liquid transactions, to the obvious consumption of foods by highly organised animals. Similar variation in degree may be noticed in the few remaining classes of activity.

To some extent this manner of description may be applied to form, in so far as its variations are not always in kind, but, as often as not, are brought to light by a discrimination of degree. But a contrast is not always drawn with enough clearness between distinctions of detail and differences of plan. In bringing some forms together, nothing in common, save material, may be apparent. Others may be found on comparison to be alike in every appearance, or differing only in insignificant details. Still, in spite of the common substance and the frequent similarity of internal arrangement, disguised though the latter may be by an outward difference of shape—in spite of similarity of organisation reducing largely the differences of plan, *there is evident a far wider range of form than of activity.*

In treating of Form we are dealing with the Externals of Life

We have already referred to the impressions of life which come to us through matter, and to the necessary incompleteness of these impressions when taken by themselves. They are an essential part of life and a necessary part of study, but they do not provide us with a complete image of the reality. The presentation of life to our intelligence in multiform shapes illustrates again, that the perception of Nature always occurs through the medium of substance. *The events of life are at one with the simplest physical changes in impressing us, necessarily, as affections of matter.* Hence we do not arrive at any wider knowledge of life, by assuming that our perception of matter and form may be illusory. The limitations of knowledge must be accepted. The means of explanation and the objects to be explained are equally relative and illusory.

The series, almost endless, of such partial images, aroused in our minds by the material of zoology and botany, is constructed for the most part of physical and chemical ideas. The phenomena of life, the form and visible change, are names for the same kind of impression as are made upon us by the phenomena of Chemistry and Physics. The conception conveyed by the word *form* is to be considered as similar in origin to the conception of carbon or oxygen, and to be treated with the same reservations.

The exchange of words belonging to the same plane of thought, the replacement of one term by another of the same degree of generalisation, does not always help the understanding, though it may render description more inclusive. The range of thought, it is plain, may thereby be enlarged, but the depth in no wise increases. Oxygen and Lungs are not in any way mutually explanatory terms. They are symbols for conceptions of the same class. The image of the vertebrate form, for example, is made up of relations *such as exist in the inorganic world*, relations which may, it is true, enter into the idea of life, but only in a subordinate and contributory fashion.

Yet, unimportant as the question of form may

appear to us when we are considering the most intimate *events* of life, it is one of the earliest abstractions drawn from the *substance* of life. Inanimate matter has, strictly speaking, no form, no constant relation to space; while form is an essential part of organic nature.[1]

Structure and Material as well as Shape to be considered

Although the word Morphology itself points to form or shape as the end of its inquiry, yet external configuration or form is often less significant than the internal disposition of the parts, i.e. *the structure*, of an organism. To a great extent, as might be expected, the form is determined by the structure, and Morphology is, therefore, mostly occupied by the problems of structure. In another of its branches, research is turned to minor differences in the appearance of the fundamental substance, protoplasm, which enable the parts of an organism to be distinguished from each other.[2]

It is clear that differences in structure might arise from a diversity in the arrangement of a

[1] It may be urged that form is part of our conception of a crystal, but, in truth, there is no real analogy. Even crystalline matter does not always exhibit its characteristic form. Inanimate matter, whether capable or not of existing in a definite form, is not like the organism in which an association with form is an essential part of its nature.

[2] The material of Histology.

material which is in itself everywhere the same, by a difference, in fact, in the pattern or disposition of similar masses. And though it frequently arises that the existence of a definite structure only shows itself in a slight modification of external appearances, there is often to be found a configuration of parts so massive and prominent as to attract notice at once. Meanwhile variation in the appearance of parts, with a consequent recognition of details of structure, is the true foundation of all that is now known about the various functions of the organism, and *gave rise to* the conception of a complex activity, wherein may be distinguished *separate processes co-operating in a single achievement.*

Moreover, it has been through the knowledge obtained from such structured objects, that we have learnt to include things which are devoid of visible structure in the class of organisms. The functions discovered, broadly speaking, through structure are found to be carried on in structureless organisms. It follows from the general participation by *the whole material* of structureless objects in the changes under consideration, that none is so readily apparent as it is in less simple organisms. A change which is carried on by a definite organ is the more easily observed and defined as the organ is more prominent, and a discrimination of certain occurrences as the *special functions* of given organs is founded upon a similar sequence of observation and inference.

It is obvious, indeed, that the extension of our ideas of complexity of function to those organisms which are devoid of structure in the ordinary meaning of the term, has arisen through the earlier recognition of function and structure in the more highly organised forms. Yet in drawing attention to this point it must not be assumed that even the simplest organism is truly homogeneous. Structureless in a sense it is, but after a time, and at a certain stage in analysis, *the conception of structure must cease to be useful.* Nearly all matter yields some clue of a difference in parts to an analysis which is carried out with perseverance.

The Value of Analysis

It is evident that the complex phenomenon of protoplasm can be resolved into component appearances. Indeed, one indication of complexity is clear enough in the chemical proof that it contains several kinds of matter. To make a partition of one phenomenon, whether it be a portion of matter or an event occurring to portions of matter, into constituent phenomena, is an operation frequently leading to scientific discovery. But we gain nothing by so doing, unless we succeed in resolving our first conception into simpler ones, or in rendering description more easy. This result follows whenever the constituents discovered are already known under other circumstances. An analysis becomes of value when it discovers common elements entering into different composite wholes.

The general results of chemical analysis afford an adequate illustration. But the ideas which we gain from analysis as to the composition of complex matter bring others in their train. A current of thought arises and takes the converse direction. When the course of subtraction or division is finished, the process may be reversed; and, what is still more important, it may be extended. *In a word, the main object of analysis is a better understanding of the conditions which regulate the combination either of substances or ideas.*

It may be noticed in other branches of research that the products of the preliminary process, whether they be material or psychological elements, can be rearranged so as to yield more than the original number of compounds. In Biology, too, the extent, and still more, the possibility, of such recombination is far wider than that of analysis. In analysing the organism into its structural elements, we may reasonably expect to obtain in biological respects as good results as chemical analysis shows in its province. But to do so, it is necessary to keep the two resultant groups of ideas quite distinct. The words *con-*

stituent and *element*, for example, when used in a biological sense, do not bear the same meaning as they do when conveying the results of a chemical analysis.

Structure demonstrates the composite Nature of vital Activity

Foremost among the difficulties connected with the earliest impression of structure is to understand how far a given kind of activity preponderates in a particular organ, and to decide whether the parts do, or do not, perform other functions, together with that which is most prominently manifested. But these are difficulties which soon disappear in a larger view of the end of function. Each distinct organ or structure in the more complex forms lends itself to some one special change, which is usually found, though not necessarily so, to be more prominent in it than others. The elements which make up the concerted action of the whole are far less capable of an ideal separation than the material parts with which they are associated. Yet, on seeking for these items of activity in the structureless organism, we discover them and learn that complexity of organisation, such as we observe in many instances, is not essential to complexity of function. The *quality* of life appears the same in every case, whatever may be the degree of organisation. The *quantity* of life, on the other hand, is something which we are not yet able to measure. The standards needed for measurement are still far from our reach.

It is an error by no means uncommon to suppose that some, at any rate, of the organs or parts of the living aggregate exhibit but one kind of activity. It is true that the easily detected differences of structure correspond with a more or less definite *localisation of function;* and indeed it is the localisation of change which renders it the more apparent to our observation. But if the whole is alive, the parts of that whole must be alive. Though the results of this life may find a different expression, or, as it were, issue in different directions under different circumstances, it is incontestable that the components are everywhere alike. It may be impossible to remove an active fragment from the organic whole without damaging it, but the needs of individuality are responsible for this.

A question naturally arises as to the manner in which the same components can give rise to different combinations, and the answer is not difficult to find. The limit of diversity in this respect can scarcely be fixed, when we consider the innumerable modes in which three or four qualities may vary in their quantitative relation to one another. Each distinct activity may be so modified with regard to the others that the possible combinations may yield an indefinite number of resultant values. *The constancy of the living elements, coexisting with diversity of combinations, has its analogy in a multiplicity of form and structure founded on a common substance.*

The Organism together with its Environment to be treated as a System undergoing Change

The perception of the organism as a material object, endowed with certain special properties and undergoing certain characteristic changes, has hitherto occupied the greater share of our attention. But necessary as it has been to the proper development of a progressive treatment of the subject, judgments made about the organism as an isolated object must soon reach their limit. The ideas which have come to us from the points of view already reached are only relatively valid. Their accuracy holds only within the boundaries which have been fixed with regard to the organism as *central and paramount*, though they may be precise and clear enough for ordinary purposes. On these preliminary perceptions wider thoughts have now to be built.

We begin to make progress and pass from the rudimentary and material aspect of the organism, as soon as we proceed to deal with its *environment*. The environment of an organism may be regarded as all *that envelope of materials and changes which surrounds it, and towards which it is responsive in its activity.* The perception of its surroundings is necessary for the observation of activity. Yet it is not easy to define the boundaries which are to be set to the environment of any given organism. They may be extremely restricted for the more minute, and indefinitely extended for the more elaborate organism. But we must begin to associate the physical state and the physical changes of the organism with the state and changes of the material which surrounds it, and endeavour to find the relations which exist between them.

In doing so, we pass from the organism as a centre of energy, influencing the material which it consumes and performing work in its attainment of that material, to consider the influence of the static and dynamic condition of the environment upon the organism. The effects to be considered are either objective or subjective: objective as observable in the nature of its activity and in the state of its organisation, and subjective as inferable from our own consciousness of sensation.

In an adequate description of any physical change it is necessary to take notice of all the objects concerned in the change. These bodies constitute what is called *the system under observation*. There may be two or more bodies composing the system, and inasmuch as every change in Nature, even the simplest, namely, *displacement*, requires at least two bodies for any observation to be possible, there can never be fewer than two items in its record or representation. Having reduced a change to its bare elements, we get greater accuracy of observation, our knowledge becomes more exact, and a conciseness in description promotes further accuracy. The device aims at the inclusion of a series of concurrent events in a single figure. The size of the system denotes the number of bodies sharing in the change, and only those bodies are regarded as within its boundaries which contribute to the condition which is finally reached.

The conception which underlies this procedure in Physics may be adopted with advantage in Biology. The organism together with so much of its material surroundings as are operative in the changes to be observed form a *living system*,[1] and it is only by including in our system all the bodies concerned that a correct estimate of living change, as a whole, can be made. The history of Physics shows clearly that incomplete accounts of physical events were generally accepted, until all the bodies concerned in them had been observed. In like case, the incomplete record of living changes is made more intelligible and more inclusive, so soon as it takes into account all the objects engaged. In either science we succeed not only in rendering our description

[1] This term must be understood here, and subsequently, as a shortened form of the phrase *the system in which life is observed*.

more faithful, but also in laying a safe foundation for more searching inquiry.

The external Changes affecting the Organism

We come to the conclusion, therefore, that our knowledge of the organism demands to be furthered by observations of the rest of the system which is concerned in its activity. And it is the changes in the environment, preceding, or occurring simultaneously with, changes in the organism, that need most attention. It may be urged that the conception of an environment is too vague for description—that its dependence upon the organism selected for observation gives the organism the prior claim to investigation. But it is precisely the connection existing between the two which renders a joint observation of them all the more important. It does not tend to exact knowledge in Biology, any more than it has done in Physics, to observe or measure one side only of a two-sided occurrence; nor is it enough to be cognisant of the existence of two aspects to the one event. The event or phenomenon in its entirety needs to be known.

The external conditions which exert a universal influence on the organism are, as might be expected, those which operate in the world irrespectively of questions of life. All matter gravitates, and the laws of gravitation are perceptible in living forms. Every detail of structure must in some degree be determined by this great physical fact. And the manner of locomotion must, obviously, be limited by it. Another property, common to all matter except gases, is *cohesion*, and the extent of the cohesive forces will fix its own special limit to shape.

In addition, the relation of the organism to *light* and *temperature* is important, inasmuch as the main material event of life, the chemical change coinciding with it, is controlled and decided by these physical factors. Light, for example, is the source of vegetable energy, and it is only within certain limits of temperature that the chemical changes subordinate to the life of protoplasm can proceed. Nor must it be forgotten that matter in the *gaseous* or *liquid state alone* is capable of rearrangement according to chemical laws, and an important limitation of vital metabolism thereby asserts itself.

To all these conditions restrictive of physical events there is a *reaction* or *response* on the part of the organism, either in change of rate of growth, or in change of one or all of the activities which qualify it for life. Other external changes which come under the heading of relations between organisms themselves, and are the antecedents of reactions such as *protection* and *competition*, are too controvertible and too involved to be treated here.

The Reaction of the Organism

The connection between form and habitat or environment, is one for which an analogy must be sought in the more direct and more easily detected relation which exists between form and function. The bond, connecting a given structure with the share of work allotted to it, is usually described as one of *fitness*. There is little to be gained from the introduction of this term, familiar as it may be, for every effort to supply it with a meaning begs the question. The means fit the end, when action takes place most directly and most economically. Such as it is, this relation connects form and function on the one hand, and the organism with its environment on the other, and necessarily so, inasmuch as *functional activity is wholly determined by its medium*. The neighbouring material, and the changes of outer origin which are incident to the organism, are physically implicated in its activity. The condition of the surroundings as well as their nature, in other words, the physical state as well as the material composition, must be taken into account

When the attention may have to be directed impartially to either side of a change which is known to be two-sided, it is convenient to distinguish the two aspects of the change by the terms *action* and *reaction*. In connection with living changes, custom has attached the term *reaction* to the organism, and the term *action* to the rest of the system.

As no outlook seems possible, wherefrom the whole biological event may be regarded in its twofold nature, we are compelled, apparently, to treat the problem in two parts. Yet it is important to remember that there is a whole, which is made up of these parts. It is also important to judge whether the whole action of the system shows a resemblance, in any of its features, to other occurrences in Nature.

In regard to one condition all changing physical systems are alike. It is found that every action manifests itself in two events, or in two groups of events, which strictly balance one another in physical value. Is there the same equivalence in the system containing the organism? Is the value in terms of matter and motion the same, whether or not we take the phenomena of the organism or the phenomena of the environment as representing the mutual change? To this question the answer is emphatically in the negative. There is not the same equivalence, though equivalence of a certain indeterminate kind there may be. *The action on the one side does not represent the same number of units of energy as are measurable in the work performed on the other, and this is the only kind of equivalence recognised in Physics.*

Some formula or other to connect the two is called for all the more, since the idea of quantitative equivalence must be discarded. One class of external change is appropriately described as beneficial. When we come to consider the reaction of the organism, which is known as a benefit, we find it is always made up of a greater or less accession of energy, and in this respect it shows a resemblance to other events. The organism, as such, is improved, and its capacity to do work increased. And this increase in energy may be manifested in the continuance of growth or development, or else in improved sensibility. But these are effects which are unknown elsewhere, and they endow the organism with its peculiarity. The existence of this reaction is continuous with life.

The conditions previously mentioned may be conceived as tending to bring about momentary and passing states of equilibrium, in an object which is itself in a state of constant instability. And here we may opportunely point out another distinction, or more correctly, another phase of the same distinction, between the living system and all others. The ordinary sequel to any physical change is a new state of equilibrium. Energy becomes in the end evenly distributed throughout the members of the system. But in the living system there is no such tendency. *The organism is constantly gaining energy at the expense of its environment, and the process of reproduction perpetuates this condition.*

It may be true, indeed, that the reaction to environment here described is a complex change made up of observable component events, or it may be more accurate to regard it as occurring simultaneously with definite chemical and physical transformations. It remains a matter for discussion, whether the subordinate changes in question, when added together, would, or would not, produce the result recognised as the complete reaction of the organism. But so much can be positively asserted, that the total change may be resolved into these elements, and that no other objective facts are perceptible. And further, it may always be said *that matter which is either chemically or physically inactive is not alive— that only what is changing is alive.*

This image of constant change, as the characteristic of life and of the individual, is already familiar, but at the moment it is especially useful, as excluding from the conception of organism all that is not undergoing change. And it assists us to realise the more effectually that much of the existing variation of form is adventitious. Form is the symbol for that which is constant in the organism, and that which persists, in spite of the change associated with it. Yet most of the living objects that exhibit a distinctive pattern owe it to the non-living matter which is associated with them. The association can scarcely be described as accidental, for in most cases this material, whether it be termed a by-product or an excretion, has a definite value as a protection or support. As integument, skeleton, or wood, it may be looked on as "functional." *In this light we may well regard specific form as only in part a true representation of vital operations, and life itself as something greater than the sum of its manifestations or instances.*

Growth and Development, like other complex Problems, give rise to Pseudo-explanations

There ought to be a very clear separation of the facts of Biology from its theories. The plain results of observation and the records of observers should be kept quite distinct from the theories and beliefs which have been based upon them. Problems arise whenever the study of animals and plants is seriously undertaken, and the more accurate the observation and the more closely facts are investigated, the more forcibly will these problems plead for any solution rather than none. But the first object is to attain as far as possible a definite knowledge of the point at which facts end and assumptions begin: to be certain, as far as one can be certain, of the distinction between that which is perceived and that which is conceived. If such an apparently simple end can be achieved, the questions which always confront the biologist will be answered less inconsequently and with fewer contradictions than is usual.

But, as we have already learnt, the facts of any science must be acquired in a certain order and in accordance with a definite plan, and we have to accept the order and plan prepared for us by authority. It is from the authoritative expression of the final outcome of thought that we derive aid in following the course along which the widest generalisations are unfolded, and in avoiding the endless divergences from the straight line of progress which are often to be traced in the historical development of a body of knowledge. An illustration of this statement will be attempted in the following description of the changes to be observed during the transformation of a small mass of protoplasm into the adult organism.

Changes, however, which are as complex as these always prove to be the most productive of imaginary explanations. It is supposed that the process of development is accounted for in some way, when its various stages have been enumerated. *There is no commoner error than to suppose that the perception of the details of a process explains the resultant changes.* It is more plausible, too, than the mode of explanation which consists in the invention of imaginary agents of change. Gravitation, as the cause of gravitational movements, or magnetism, as the cause of magnetic phenomena, are examples of such fictitious agents.

The same kind of mental defect leads to the widespread fault of assuming, when a constant association of events is perceived, that the one explains the other. An association of events may be one of sequence or one of coexistence, and in either case a quantitative relation may be manifested in addition. The admission of an association as being a necessary one will follow a proof, that suppression or variation of one event coincides with the absence or variation of the other. But when this has been described there is nothing more to be said. We are still as far away as ever from understanding why this or that association should exist. In the exceptionally complex relations of the organism, it is unusually easy for descriptions of this kind to be regarded in the light of explanations, one event being selected in that case as the so-called *cause*, and another as the *effect*.

It may be said that the starting-point for every organism, whether plant or animal, is a single cell. But certain difficulties arise the moment we try to make this statement hold as a generalisation to include the origin of every form of life. A protozoon, amœba for example, is both a single cell and an organism. It divides and two amœbæ exist. The single cell has given rise to two organisms by disintegration. Again, in more complex forms of life the starting-point is not a single cell, but two cells fused together. But there will be no misunderstanding, if a general description of what is most widely observed takes the form of stating, *that an organism changes from singleness and simplicity to manifoldness and complexity in the course of time.*

The beginning, then, of the process of development, so far as it is observable, lies in a simple mass of protoplasm, and at the end of the process we may have a structure of very diverse components. Between the beginning and the end lie the numerous steps of a manifestation, concerning which conflicting hypotheses have been constructed, and the most varied opinions have been held. But the project now before us is not to consider the hypothetic causes of the changes, but the changes themselves as they appear to us in their most prominent features. We cannot do more than watch the gradual accumulation of matter around the primary cell, observe that life spreads over each successive increment, and trace the progressive differentiation of parts. Herein is encountered a phenomenon resembling that of specific form, *but now it lies concentrated in the individual.*

Cell-division is an Incident of Growth and of Reproduction

Now the most marked change which the cell

exhibits is its spontaneous division. The same fact would have taken a more familiar form of expression, if the words *power of dividing* had been used instead of the word *division*. But familiar phrases are not always correct, and in stating a fact of such importance, one, indeed, of the primary facts of Biology, words must be carefully chosen. What is observed, under the most varied conditions and in all kinds of living cells, is either the process, or the result of the process, of division. The process varies considerably, and the end of the process differs under different circumstances.

Now, whatever the intermediate changes may have been, the appearance of two or more living objects, where one of the kind has been previously apparent, is the fact of reproduction; and it makes no difference whether the living object is a cell or a group of cells. The character of reproduction is repetition or increase in number. In spite of all efforts, we can do little more than trace out the barest outlines of a relation between parent and offspring. All that we can observe is a final increase in the number of individual objects, and this we can trace to a division of matter and a more or less complete isolation of material parts.

And this is cell-division. It may be taken in detail. Rearrangements of matter occur in a cell, and its multiplication ensues under favourable conditions—that is, an internal boundary appears, separating one portion from another. A large and complex object may be observed to multiply, and the larger operation may, in turn, be traced to a progressive accumulation of smaller changes such as the cell exhibits. It may be said that the innumerable cells, of which a complex animal is made up, have been produced by the division, first of all of the ovum, and then of its products, and so on for each group of multiplied products.

In this description, both the increase in the size of an individual and the increase in the number of individuals are referred to a process of cell-division. Both growth and reproduction, which is a kind of growth, or rather an extension of growth arising from isolation, are changes in outer dimension or form, taking place simultaneously with internal reproductory events. There is nothing in the internal changes, then, but a repetition on a small scale of what is observable on a large scale. The processes may go on indefinitely, until death checks the number of individuals, and the approach of death checks the increase in the mass of the individual. *Death is the sequel to reproduction, both for the group of cells forming the organism and for the single members of the group.*

A question naturally arises as to the criterion by which we judge of the creation of two individuals from one. What determines the distinctness or separate individuality of the two? If we were always confronted by an isolation in space, the answer would be simpler. But in the minor reproductory changes of growth, and in cases of colonial forms of life, there is no *complete distinction of locality.* A separation by means of a limiting membrane, either cell-wall or integument, it is true, may frequently serve to identify individuality. But we find a better answer when we turn away from the morphological presentation of life. *If we consider the activity of the organism, we can perceive in the complete independence of reaction towards environment a genuine case of reproduction, and in the partial independence of such reaction the ordinary course of growth.*

The most prominent and most general activity of living forms is growth. Every other mode of activity which they display may be traced directly or indirectly to it. Even sensation demands, for its continued existence, a growth to replace the nervous waste which it entails. It is a fundamental property of the organism to grow, inasmuch as every change which takes place in it implies a loss of substance. In part or as a whole, it must die, unless its losses be constantly renewed. By cell-division and reproduction of parts the organism is maintained. By separation of parts, and by rapid and energetic growth of these detached parts, the species is maintained.

The conception of growth as an incessant change of constant character is an abstraction which covers the whole of life and likewise the continuity of life.

In the equivalence of all growth to a process of multiplication which is universally existent, except so far as it is checked by death, we recognise the origin of an important group of secondary conditions or factors of life which exercises great influence on the organism. Inasmuch as the supply of food is limited, an increase of numbers sets in motion those conflicting changes within the living world which affect many members of it so profoundly. The competition for food, and in many instances the danger of being utilised as food, bring into the environment of each organism a further complication of events. The typical living system does not always consist of an organism in a completely inanimate world; it may have to include and account for activities between organisms themselves, and it is impressions from these which have been most productive in controversy and speculation.

The Idea of the Cell to be used with Reservations

The useful conception of the cell is derived from a process of analysis, and the result is of the same quality as the object analysed. We cannot expect to find in it the explanation of the organism. The component cells are of the same nature as the aggregates which are constructed of them. Those who seek in the parts for a solution of the whole will fail in their search. Every cell is, in a certain degree, an organism, and cannot serve to explain that which it resembles. Those who take the cell as an instrument for order and arrangement will employ it most profitably, as once it is made the analytic unit, there is an almost endless series of reconstructions to follow. Aggregation and modification of cells then become a mental framework, upon which we may arrange the innumerable variations of living form and structure.

To disregard the part which has been played in the methodic arrangement of our thoughts by the theory of the cellular constitution of plants and animals would be illogical. It is a theory which has been fruitful in many additions to our knowledge, and if it no longer gives promise of such important discoveries as have been associated with it in the past, it still remains a valuable aid to description, and, in consequence, an important instrument in adding clearness and definition to our thoughts on form. When the statement is made that plants and animals possess a cellular structure, it must be remembered that this is not always apparent; and when apparent, it should be looked on as a basis for description rather than a source of explanation. The course of investigation for a long period took the cell as its objective. It has more recently turned in the direction of the nucleus. The want of a definite cellular structure in Mycetozoa, combined with the presence of numerous nuclei, justifies the transfer of the attention from the cell to the nucleus. Yet by analogy with the past the nucleus, too, must give way in its turn, and the material seat of life show itself in structures still more deeply hidden.

The Products of Cell-division not necessarily alike

But there are certain matters of observation to be taken in conjunction with the fact of cell-division, before its results can be properly estimated. Cells may vary very greatly in appearance. In vertebrate animals, for example, great diversity is shown. The existence of structure implies such diversity. In animals or plants of simpler structure there will be less variation in the appearance of cells. But if the cell reproduces its like, just as the organism does, how does this difference arise? But neither cell nor organism does always produce exactly its like, though the reproduced organism in its totality differs by no essential from its forbears. And it is at this stage that our descriptions, hitherto parallel both in gross and in detail, must begin to diverge.

In the adult state a muscle-cell reproduces muscle-cells, and nerve-cells are the ancestors of their fellows. In the same way the organism ultimately produces its kind. But on the other hand, that which is the origin of every cell in a given organism, and of the organism itself— *the germ-cell*—does not merely reproduce its kind. *It reproduces, by slow and continuous variations, every kind that is known.* The limits are the known differences of form and composition in any given organism, and these limits have been reached in the interval of time which has been occupied by that organism in reaching maturity, an interval which is in many cases remarkably short. *Cell-division in the mature organism has a result which is different from cell-division in the embryo.*

Most forms appear first as objects which are to all investigation the same in structure and substance. It is impossible to judge from observation of the ovum the course which development will take, or of what organism it is the antecedent; yet differences of structure may still lie hidden beneath that superficial uniformity. But it need not be imagined that the final difficulty would disappear with the detection of such hidden differences.

The first cell divides. A line of division, more or less apparent to the eye, comes into existence. But it is difficult to say how far this change betokens independent activity for the parts, merely separated as they are, by a boundary which is nothing but slightly changed protoplasm. Further division of the same nature ensues, and with increasing assimilation the number of such structures reaches an incalculable value.

But while this multiplication goes on, a change of form is proceeding, and a scheme of arrangement coming into existence. This is the process of development. Groups within groups are formed, and activities in keeping with the variations of form and arrangement appear. And such complexity of structure as may appear is not wholly indebted to variety in its ultimate components, for these are often closely alike. It is more directly traced to variety of arrangement. Yet, though the stages of that process which gives rise to the varied appearance of cells are little known, the effects have a prominent diversity. It is an advantage in the arrangement of the numerous facts which meet us at this juncture, to regard growth and development as distinct processes: growth as simple repetition of the like, and development as the effect of an external agent of change modifying the effects of growth. Yet they are both contained within the single conception of a *growth which must be, in accordance with its nature, differential in its results*, unless circumstances are too strong and some influence of locality or constraint of position induces uniformity.

But though the antecedents of growth are so

little known, the results are plain; and the terms, growth and development, are convenient as a description of two component changes, which together bring about the existence of structure in the individual and likewise of diversity in the living world. The combination of processes known as development, manifests itself ultimately by the appearance of a certain symmetry in a structure and outline more or less involved, and this appearance is what is recognised as the organic form. In other words, the results of development show themselves both internally and externally, and the fact of varying structure corresponds with the fact of species.

This double result of a differentiating process enables us to treat the progressive stages of structural growth in the individual as parallel with the diversity of specific form. The differences *at the various stages of individual growth* can be very easily brought into line with the variety of form exhibited by *distinct species*. The material continuity of one series separates it clearly enough from that in which there is no connecting link, save a vague genetic relationship; but our outlook on either series is alike in being taken from the same intellectual level. The comparison is useful, and it may be made without entailing a reference to the wider hypothesis of *evolution*, which does not come into our scheme of discussion.

But whatever may be our attitude towards the important principle of evolution, we must admit in it the attempt to bring all the phenomena of life under the rule of a single law, and to give Biology a shape to which every body of knowledge must at some stage conform. At the same time it is clear, that the last and most complete generalisation of physical life to be gained from existing means of practical observation is contained in the idea of differential growth or development. To go beyond that statement is to enter the region of speculation.

Prejudice with regard to Form and Change

The customary view of the results of growth, just as of vital changes in general, is based upon the nearest and most familiar examples. It is habitual with us to regard that which is familiar as normal. The modes in which functions are performed, the apparatus of the senses, and the whole elaborate structure connected with vital activity, as gathered from an examination of man or other vertebrates, are apt to bias our judgment about other organisms. Either as a whole or in detail it is difficult to get an accurate conception of the *possible* manifestations of life. Yet, to give an example, an eye is not necessarily an organ like our own. It may be a slightly changed portion of the exterior. Nor, again, is it in any way unnatural for an ear to be borne upon an organ of locomotion.

Then as regards the individual at successive periods of its life, we must not always expect to find a graduated and continuous progression to a maturity. So sudden, indeed, are the occasional breaks in a given course of growth, that the different *stages* have been, and reasonably so, regarded as representing different *individuals*. Such phenomena may well prompt us to reconstruct our conception of the individual. We cannot consider the idea of constancy of form, or of the progressive attainment of a constant form, as an essential part of our conception of the individual, in face of the numerous instances of *metamorphosis* and *alternating generations*.

The action, too, of the organism in assimilating matter which is attainable and suitable, one of our earliest observations, is a change which can be made to bear a different meaning, as soon as we rid ourselves of the associations connected with the word *food*. If we regard the material universe as divided into two main classes of matter, the one alive and the other inanimate, it ceases then to be a question of food, but rather of *the shifting locality of life*. Now one portion of matter, now another is touched by it. Such a view enables us to take a better grasp of that which is more important than the material itself, namely, the dynamic value of the change occurring in it. The potential activity of food-stuff is wanted, not the substance. Or again, it may be necessary to regard inanimate matter as liable to become alive, to come in the path of that special transference of energy which proceeds slowly through an indefinite quantity of matter. Any change in our outlook on these events is better than standing still.

Attention to the Environment assists in Comprehension of Development

With less prejudice respecting form it would be easier to generalise about its diversity. Now we succeed best, and the widest generalisation about form, as about other facts, lies before us, when we include both the organism and its environment in our regard. It is not difficult to perceive that a mutual relation exists between the nature of the environment and the living apparatus which reacts to it. Yet it must not be hastily assumed that the relation interprets either *itself* or the *facts which are related*.

A larger and more varied environment corresponds with greater complexity in the structure of the reactionary agent. Then the more varied the possibilities of equilibrium with a Nature divisible into endless environments, the greater will be the number of species in existence. Here, too, we perceive a parallel between diverse species and the phases of the individual. And the existence of the parallel will show more clearly in what is subsequently described.

The existence of separate continuous combinations of definite changes, localised in different individuals and in different species, is a form of expression in which innumerable fragments of observation are recorded. But the account rendered by these words is still far from adequate. Frequent emphasis has been laid upon the self-controlling and self-originating function of the organism, which sets upon it a mark of distinction from every other changing object. *The changes in it, which are associated with surrounding events, are in truth adjusted by the organism itself, to an end which can only be described as the advantage of the organism.*

The exercise of selection in the matter of food has already been mentioned, and we now come in touch with a capacity to adjust or adapt, that is, to select, the total activity. The *mode of co-ordination* selected with regard to external changes is one and the same thing as the *kind of individual*. And a process which is undoubtedly spontaneous, in some degree at any rate, ends in the existence of a specific form, and records itself therein. *For self-originated variations in the relation to outer changes, if they persist, leave their impression in modifications of form.*

Extension of Idea of Environment

The effect of surroundings is not confined to the organism as a whole. It extends to each definite part or organ, and even to each cellular constituent; and in just the same way as we may regard the activity of the whole as conditioned to some extent by locality, so may the function of an organ or cell show plain signs of being determined by its neighbours as well as its antecedents. In this case the principle of environment is pushed to its logical limit. But the same idea may be clothed in the statement, that the conception of an organism is extended legitimately *to include a cell*, even when the latter is subordinating its own activity to that of the aggregate. And intermediate between the whole aggregate and its ultimate cellular elements, we have minor combinations, or organs, subject to influences which are similar in character.

The influence of the environment upon its reagent is illustrated in almost every detail of Morphology, and one example may be quoted. It is observed that protoplasm can only grow by an extension of its surface, that is, of the boundary separating it from its surroundings. A general process of condensation is unknown in growth. *And the need of an extension of surface confirms the existence of a mutual action between living matter and its environment.* This requirement can be secured by an even expansion in all directions—ordinary voluminal expansion. But a limit is soon reached in this way. The ratio of total mass to surface may become too great to permit the necessary exchange of material between the outside and the interior of the organism. Sufficient surface may, however, be provided by an alteration of shape; and there are two classes of alteration, the one in which *internal cavities and channels* are observable, as in animals, and the other in which a *branching extension* is seen, as in trees. By either method, *the minimum of distance of any portion of protoplasm from its correlated environment is secured*, and we may recognise a further association in the fact that the food of the former is solid and requires solution, while that of the latter is fluid.

Two main Paths of Development

There still remains a difficulty to be surmounted, in connection with the complex relations existing between the converse parts of the living system. The persistence of life itself we know to be guarded by reproduction. Yet no consideration has been paid hitherto to the persistence of species in reproduction. It may be answered that the same environment envelops both parent and offspring, and moulds them alike. But, on the other hand, it is characteristic of the organism that it is able itself to initiate changes in its relation to externals, and in view of this capacity we must say, that it can, within limits, *vary its own environment.* An increase of susceptibility corresponds with an enlargement of horizon.

We are face to face, then, with a dilemma. Does the environment determine the organism, or the organism select its environment? Perhaps the best answer is to say that both are true, and that each reacts on the other indefinitely. *Nowhere do we see the organism free from external control, and, so far, the environment for each species is rigid; yet innumerable are the adjustments whereby the organism escapes from control, and proves itself more constant than its environment.* Even purely physical life, then, cannot be truly described as being wholly what its environment makes it.

For the present, then, the conception of species takes the shape of a series of resting-places, in a continuous effort of living matter to arrive at an equilibrium with inanimate nature, and these we find ranging in progressive stages from the simplest unicellular organism to man. And the progress may either take the direction of accumulating the maximum of energy by means of growth, whereby the organism incorporates and assimilates as much as possible of its environment; or it may branch off in the direction of increasing nervous development towards a maximum of sensation or susceptibility to environment, a development which, in one sense, is passive and reactionary, but, in another, tends to bring the whole of Nature within the increasingly active perception of the organism.

Hypotheses useful in co-ordinating Facts

We are again met by the reflection that *observation*, however accurately performed, does not add to our *comprehension* of Nature, unless it is controlled by a scheme of thought for which confirmation or disproof is sought in the results of the observation. Some principle already shaped, whether it be a product of our own mental activity or be handed on to us by authority, is wanted to give to each fresh item of perception a rank in an orderly plan, and to add clearness and value to that which would be, otherwise, unrelated and imperfect. The inability to grasp more than a certain number of facts without the aid of theory has already been demonstrated by the course of this inquiry. *Beliefs, hypotheses,* and *theories* have forced themselves upon us.

In the acquisition of each science, the same feature is prominent. Successive stages of progress are reached through a series of changes in our system of conceptions, quite as much as by extending our observation. The arrangement of different kinds of matter in an intelligible manner did not begin until the *indestructibility* of matter was perceived. The classification of bodies according to composition is based upon a belief in the existence of *elementary substances*, which never change in total quantity nor in kind, however varied may be the modes of combination and association of parts of these substances with one another. A science of Chemistry is created, wherein all forms of matter exhibit a relationship in respect to the number and relative quantity of their components, and also to the conditions under which they have entered into union.

In Physics again, every change is brought into relationship with the great generalisation of *the conservation of energy*. Whatever may be its nature, a physical change first takes a meaning when it is regarded as an example of a certain quantity of energy being transformed without loss, and it is intelligible to us only in the light of the principle that energy is indestructible.

It is only natural that in a science of later growth dealing with the most complex phenomena of all, the most general principles should be still in the stage of controversy. *There is, so far, no law admitting of universal application to the subject-matter of Biology*, though more than one generalisation has assumed that position, and rendered temporary service in so doing. Hypotheses, which are only partly true, may be of value in the classification of facts; but if progress continues, there comes a time when such partial truths must be absorbed in wider conceptions. In the familiar image of an organism as made up of a variety of living units or cells, we are able to discern many imperfections, in spite of the unquestionable value it still maintains as an incentive to research.

Ideas of Cell-arrangement more useful than Ideas of Cell

The perception of a cellular structure in itself avails little. It is from theories of the arrangement and association of cells that the science has gained more than can be easily estimated. In the plasmodium, an accumulation of units presents to us nothing more than is revealed by the single unit. Volvox exhibits the simplest illustration of a difference in the form and activity of individual cells when combined in a group. The associated units begin to differ in their function, and they grow in successive stages of complexity to differ still more widely. And it is this conception of development, that is, of divergence in the subordinate parts of a whole, which confers upon the cell-theory its value as one of the fundamental principles of Biology. *In fact the discrimination between a plasmodium and Volvox marks a new epoch.*

Arrangement of Cells less significant than the Combination of Functions

In the earlier days of botanical discovery attention was mostly directed to the cell-wall, and only after a lapse of a hundred and fifty years was it discerned that the *cell-contents* give rise to the cell-wall, and, therefore, signify more in the life of the plant. And now, at the present day, it is more and more impressed on us, that after a certain stage in comparison is reached, the most suggestive distinctions between organisms are *physiological* rather than *morphological*. It is appropriately considered to be not so much a question of degrees in the accumulative grouping of cells, as of *permutations in certain dynamic components*, which combine variously to produce a common end, namely, the resultant activity of life. *In varying degrees and in different capacities each unit of activity takes part in a larger scheme of change, which demands for its recognition a conception of life as a whole in relation to the organism as an individual.* And yet it may be pointed out, that just as subordinate relations may be readily detected underneath the bolder outlines in a comparison of form, whenever a certain degree of complexity has been reached, so may it be noticed that details of structure often perform, with apparent fortuitousness, secondary functions, and do so, too, with advantage to the organism. The capillary system, the primary function of which is the distribution of food-stuffs, affords an illustration of this point, when it serves to regulate the temperature of the body.

In transferring the attention from the grouping

of the cells to the mode of combination evident in function, we leave the problem of specific form without a solution. Though the activity of each cell may be merged in the activity of the whole organism, yet there is no fundamental difference in most respects between the complex group and the single cell; and, to adopt a broad description, *the organism may be said to live as a cell lives.* But to this statement one notable exception must be made. The correspondence between the solitary cell and a system of cells does not extend to that group of occurrences which is connected with the state of *nervous development.* The function of nervous organs is not easily apprehended. It is a phenomenon, peculiarly indeterminate and yet peculiarly significant, which is often described as the susceptibility of an organism to its environment, or of its parts to one another.

Nervous Development connected with the highest Function

The circumstances of all modes of vital activity, or in other words, the conditions under which they exist, constitute their environment; and in the general search for what are called causes, it is natural to look to the environment for some explanation not only of function, but of the form of living objects.

With regard to the structural plan, there are

many signs of the condition which mechanical experience prompts us, as already stated, to describe as suitable or fit. The forms of birds and fishes, for example, are *fitted* respectively for locomotion in an aerial or liquid medium. But it is the *diversity coexisting with similarity of environment* which calls for explanation, and which has brought into existence such hypotheses as adaptation and natural selection.

But the reluctance which may be felt, in considering the constitution and condition of the environment as a sufficient cause of specific variation, will be increased when it is remembered how seldom the conditions of life alter, except as regards the range of environment. And even this extension cannot affect the greater part of the activities. A larger quantity of food, for example, is not necessarily assimilated.

It is habitual with us to regard a given change or activity as possessing magnitude in virtue of a rate or extent. The change may be more or less intense, or it may be distributed over a larger or smaller area. The material in action may possess a large quantity of energy in comparison with its bulk, or the magnitude of the energy may be due to a large quantity of material finding itself in the same condition of change.

But any comparison of this kind fails to give satisfaction, when we come to deal with the question of the unit of life—the individual. This fact does not suffer its measure to be taken in any way. Its position in a scheme of cognition is determined by other characters. *And among the characters which help to make up our knowledge of the complete individual neither quantity nor intensity of action finds a place.* It is, on the contrary, a balance of activities or a consensus of changes which has to be discovered in the totality of every organism.

On the other hand, the range of sensibility is obviously equivalent to the extent of the region in which it is exercised. But the complexity of the environment depends upon the degree of nervous organisation, and though sensation itself is of necessity consequent on outer changes, its development is of independent origin. *The chief cause of diversity in species may be traced to the organism itself, though not to its substance; for it is noteworthy that the development of the nervous system carries with it many subordinate changes of organisation, as well as a wider range of function.*

In the lower animals the centres of susceptibility are fairly evenly distributed throughout the form, while in the simplest organisms each separate cell appears to react more or less independently. As we traverse the range of animal life, however, we find the nervous system becoming more distinct and more concentrated.

The chief ganglia of the worms and medusæ, to take an example of the first case, are distributed either segmentally or in the form of a ring, so that there are a certain number of parts all supplied with similar central nervous systems, though the latter may be connected with one another. And in the existence of any such special system we have the first and last problem of physical life before us, whether we fix our attention on its sensory or its co-ordinating function.

Internal factors of change of form must be accepted, and yet external factors are not to be ignored. However true it may be that the antecedents of variation reside within the organism, and whatever may be the indications that development is a constant accompaniment of all growth, there can be no doubt that there are, at the same time, external factors active in checking and limiting the results of the impetus starting from within.

Equilibrium checked by the Organism

Again, deeply-rooted differences are apparent in the relations of living changes to the universal manifestation of change, which we have learnt to look upon as indications of an objective reality— *energy*. There is ample evidence of a tendency to equilibrium in the case of exchanges of energy among inert bodies. The *degradation of energy* is an invariable accompaniment of inanimate changes. A material system absorbs energy in such a manner that its usefulness is diminished. It becomes evenly diffused throughout the whole, and is lowered consequently in *availability*. This process of diffusion extends to the world at large.

On the other hand, every living organism is a centre at which this action is resisted. Yet this statement in its bare form may not convey an accurate impression of the whole truth. Though equilibrium is not general, there is, between animal life as a whole and plant life as a whole, a rough balance as regards energy. The total changes in energy in the one case balance those in the other. A scheme of *general symbiosis* exists comparable with that which occurs with two symbiotic individuals, for instance, when an alga and a fungus exhibit joint life as a lichen. *The partition of such changes into isolated groups and localities creates the individual organism. And wherever a break in the process of attaining equilibrium occurs, there an individual lives.*

Summary

Whatever be the point of view, life will be found resolvable into component activities, and structures will be found in correspondence with them. Yet all of them, activities and structures alike, will be found, however detailed the analysis, to possess the character of subordinates in a scheme rather than the quality of independence. No facts are more surely impressive than the agreement in action and the consistency in result. And at every point of observation a contrast with the rest of Nature presents itself. At each centre where energy is transformed and where the phenomena of the organism appear, the course of inanimate events is always modified and frequently reversed. And, moreover, it may be observed that there is among the qualities of the organism an ability to perpetuate this state of affairs through the incidence of reproduction.

Yet when we remember, that it is by way of assimilation and the attendant event of locomotion that this energy-exchange takes place, and that assimilation is but another name for the occurrence, we perceive how incomplete is the broadest view of physical life, and how much is left for consideration when material conceptions are exhausted. Assimilation is truly an early, yet essential, stage in a process which ends, in the higher animals at any rate, in perception, though the provision of material for assimilation is dependent on the exercise of that perception. There is evident, in the relation of the organism to its environment, a qualification of the phenomena of life which may serve to define them.

Furthermore, while the reactions between two

FORM AND DEVELOPMENT 227

portions of inanimate matter are such as can be included in all cases under physical formulæ, with the advent of life this no longer holds; and it is this fact which gives to the science of Biology its distinctive character. It is true that living matter is likewise under the operation of physical laws, but it is the constant evasion and counteraction of those laws which mark the existence of life. The possibility of recognising life, except from expressions of feeling by a sentient being, would otherwise be impossible.

The remarkable constancy of the living form, one of its distinctive signs, even when united in thought with the ceaseless occurrences tending to disturb that form, gives no positive indication of other than physical agents. Indeed, it is only by a just apprehension of everything that concerns or affects the organism, in other words, by a due regard to external changes as well as to the more prominent activity of the organism, that it is possible to gain coherent knowledge of the fact known as life. When the persistence of a state of dynamic equivalence between the organism and its surroundings has once been recognised, either in the continuity of individual life or in the fact of reproduction, we begin to discriminate between life and ordinary physical change, and to realise that life itself is neither wholly material nor wholly phenomenal. Then, at any rate, will living objects be recognised as a class by themselves, apart and distinct, even though the mystery enveloping each individual organism may be conceded to extend to the activity which is displayed by Nature as a whole.

CHAPTER XIII

THE MEANING OF SENSATION

Introduction

THE apposition of the organism to its environment has already been described in one of its aspects. A broad outline has been given of a certain varied group of events, which may be said to approach in some of their features the changes manifested by unorganised matter. One axiom in particular is found to record a circumstance common to all action, whether connected with life or not, that no change, however simple, can be fully mastered by watching one body alone. Even the minute and intricate occurrences which are originated and completed within the organism itself come under the same principle of mutual dependence as obtains with those of a larger operation. In dealing with the living system in this connection, we came to the conclusion, that in the organism characteristic changes originate and are made subservient to the sum of its relations to an environment. And the sum of relations existing between the parts of the living system is not a mere correspondence between two groups of changes, one of which is within the organism and the other external to it.

On the contrary, if we have not already gathered from our inquiry a clear conception of the organism, we have learnt one thing, at least, *that it is not a mechanism of fixed invariable parts*, nor a system of organs with definite functions, but rather an intricate and co-ordinated *group of self-regulated changes*, which are located in an unstable, evanescent, and recreating substance. And this substance is responsive to external changes in a manner which can only be described as beneficial to itself, and with variations in degree which are too subtle to be reduced to law, or to be reached by quantitative calculation.

Now locomotion by means of a special muscular structure, the assimilation of external objects and many other events, are objective incidents, more or less apparent to superficial observation and plainly obedient to the laws of energy. But there is no observable process whereby the organism can manifest itself as a subject. Indeed, the terms—*process* and *manifestation*—are not legitimate in connection with the idea of the subject. From inference and judgment something may be learnt about it, but not from observation.

In an attempt to form a conception of the subject, it is necessary to consider, in the first place, the various objective activities of the organism as uniting in a single resultant individual. A union or consensus of changes in the individual produces *a combined reaction to the environment.* After we have abstracted as far as possible that which is constant, namely, *the individuality*, from its characteristic envelope of persistent change, we may proceed to the further task of inferring that this unitary object is likewise a subject— that there is, so to speak, another side from which it may be contemplated.

In the discussion of Sensation it will be necessary to deal with this dual aspect of the organism. The difficulties which surround it are sufficient to explain the postponement of that discussion to this stage of our course. The ideas which are still lacking in our conception of the organism do not appear till late.

Illustration of ordinary mutual Action

One of the simplest changes in which two bodies are concerned is that which arises from their impact. If the movements of two bodies are such that collision occurs, either a redistribution of those movements takes place, or the bodies may come to rest with regard to each other while suffering a change of shape or state. The mutual action brought about through their collision is rendered evident by a change in the bodies as wholes, or in parts of the bodies. By investigations of these physical occurrences we have learnt one thing about the impact of elastic bodies of known mass, *that a simple formula can always represent the relation as regards quantity existing between their initial and final motions.* And even in the case of bodies which are not elastic, it is confidently believed that the occurrences, if they were all observed, would be found to come within the range of the law of the conservation of energy. The results may be, in fact, quantitatively foretold, if the initial states and properties are closely observed.

The end of all dynamical or physical observation is to arrive at such a knowledge of the mode in which mutual changes occur, that we can foretell what will occur under any possible circumstances. By adequate observation the general law, according to which given changes occur (or, in other words, the abstraction from all observed cases of a general expression or formula) may be utilised *to anticipate what will take place under new combinations.* And indeed the possible products of all physical activities—the work to be done by physical agents—can be foretold in most cases with extreme accuracy. The results obey so closely the quantitative and qualitative conditions ascertained by previous inquiry, *that we consider it the same thing to know the circumstances of a purely physical event as to know the event itself.*

In fact, according to a well-known generalisation, every change which comes within the scope of Physics may be expressed in terms of *matter* and *motion*. The antecedents as well as the results can be measured, either directly or indirectly, by standards of length, mass, and time. Moreover, the sum of the numerical values obtained by the measurement of the action is equal to that obtained from the measurement of the reaction. This is the meaning of the principle of Conservation of Energy. We have an illustration in the foregoing statements of the standpoint of Physics with regard to Nature, and of the end to which all physical inquiry tends.

Additional Results when one Object is alive

When we come to regard the contact of a body with a living object, we are confronted, as a rule, with a complication of results which cannot be reduced to any *definite law*, much less to one of quantity. That which happens is neither certain nor constant. Even in the case of so simple an occurrence as contact, no one would venture to assert that the result is dependent upon mass or other physical properties. Every physical agent produces, under similar conditions, similar effects. Moreover, it *must* produce similar effects. Nothing can be added to, or taken away from, the full effects as laid down in the provisions of a general law. In the case of living objects, on the contrary, anticipation of results can only take place within a wide circle of probability.

If we fix our attention on any of those changes which make up the body of physiological knowledge and associate themselves with the facts of Morphology, we find a wide range of variety in their mode of operation and in their conclusion. And the range in question affords opportunity for the exercise of that selective control which is a special feature in the activity of the organism. Yet since the material basis of life is what it is, gathered from *the same store* as supplies the medium for all the events of which we have knowledge, there is no cause for surprise at the existence of some limits to the mode of action shown by the organism. It *selects* this or that method of attaining an end, but it does not override or annul universal physical laws. That which is special in the organism is *additional to* the conditions which obtain for all objects.

The antecedent of every sensation is an act of contact either with ordinary matter or ether; and in all cases, except *touch* and the *muscular sense* (of *weight* and *inertia*), it is a disturbance of a vibratory kind which produces sensation. Perhaps in touch also the underlying stimulus is a molecular vibration. At the surface of the body there are receptive organs connected by nerve-fibres with a central nerve-system. The latter may be described as *the seat of a physical metamorphosis*, in much the same way as the substance of the organism in general may be described as *the seat of its life*. The nervous message, whatever form it may take, is propagated with many accompanying circumstances, and with a definite ascertainable speed, along a clearly defined nervous route.

There are, therefore, three distinct stages in the event of sensation—an outer material change, an inner nervous disturbance, and, lastly, the consciousness of sensation itself. *And they are all different in kind.* The outer changes or *stimuli* do not resemble in any respect the process of nervous transmission immediately preceding the sensation, although the transmission is known to be mainly made up of physical elements. Hence we are face to face with two remarkable events of transformation, one at the surface of the organism, and the other in the interior, before the organism is conscious of its environment.

But in a highly-organised animal, such as must here serve for illustration, we have the widest possible opportunities of learning about sensory processes. A question may well be asked as to the relation which exists between the environment and an organism of less structural complexity. And, still further, by what means is the *plant*, or the *animal without nervous development*, put in relation with the external world? When a phenomenon is so incomprehensible even after analysis, the unanalysed event must prove still more mysterious.

Effects due to external Things belong to two Classes

In returning to the important conception of the organism as something peculiarly susceptible to external change, it must not be forgotten that there are *two distinct classes* of change possible in the system of which the organism is a member. To the one class belong material modifications which are manifested under similar conditions *by organism and environment alike;* to the other belong those subjective events which the *organism alone* can experience, though the environment has a share in producing them. The effect of light or pressure upon plants in altering their shape, the modification due to the physical properties of the habitat, perceptible, for example, in the form of fishes, will illustrate sufficiently the effects which belong to the former class.

No one would entertain the suggestion that consciousness accompanies either the *more immediate change* displayed by the plant, or the *slow progressive reaction* which, after many ages, has reached its final expression in the fish-shape. Nor yet, indeed, should there be surprise at finding, in the operation of these influences, many results which are analogous to, and even closely correspond with, other physical consequences. Certain of the effects of pressure, gravitation, light, and other agents, are naturally alike in both animate and inanimate objects.

But the class of effects which is now specially engaging our attention, though the antecedents are the same in character as before, stands by itself as without parallel in the inanimate world. Yet care must be exercised lest this statement be burdened with an implication, that all living matter is endowed with sensation as we ourselves know it. Susceptible to outer changes, and sensitive in a fashion, we observe it to be. But the terms—*smell, sight, taste, touch,* and *hearing*—owe too much to association with other ideas in our own minds to be applied indiscriminately to all organisms.

Into a link between living and inanimate matter, sensation has been mentally fashioned; and it may be said with some truth, that it is *the link itself* which animates the organism, and that sensation is the dominant character of life. But not necessarily do all organisms *see* and *hear*. Sensation is not likely to be always such as comes into the experience of the most highly-developed creatures. Variety of form alone would point to the improbability of similarity, if the variation and even absence of specific organs did not render it still less probable. In many cases there is no apparent mechanism which acts, as our own sensory apparatus does, to mediate between the organism and external change. Yet the rejection of innutritious particles by an *Amœba*, or the movement by which *Drosera* captures a fly, exemplifies in all probability the existence of a relation, which finds its highest expression in man by means of one or more complicated nervous occurrences.

No mechanism appears in many cases, but the same kind of difficulty continues to surround the events. The existence of intermediate changes in the more complex forms gives no assistance to the intelligence. Perception of external objects and selection of food are traits exhibited by the most minute organisms, and there is no reason to consider plants entirely devoid of them. *The reality underneath the term sensation is something belonging to all organisms, and in all cases it eludes our comprehension equally.*

The Character of Sensation

Of one thing we may be certain, that the effect produced upon living matter by external objects and changes has in all cases something in common, something by which we may distinguish living objects from those which are inert. The predominant idea is undoubtedly that of impression, as of a soft body in contact with a hard one, and of different kinds of impressions or marks being made by varying stimuli on a yielding organism. And in this connection the organism is the passive recipient of external influences. Contact and touch are obviously the same; also, but less obviously, hearing, sight, and smell are due to contact of some part of a living surface with substance[1] in motion. What is observed when such contact, whatever its kind may be, is made with the organism? Sometimes a *movement*, bearing necessarily no definite relation to the disturbing agent, takes place. Sometimes there occurs an alteration which is not perceptible from outside, and only distinguishable by elaborate research. And all the various modes of change which can be summed up in the description that the organism is sensitive or excitable belong to one of two classes—perceptible physical movements or imperceptible subjective sensations.

On the other hand, we must remember that all objects are sensible in a certain manner. They receive impressions and are changed by them. Many illustrations of this statement are familiar, among others, the exchanges of motion and energy among inanimate bodies, the effect of light on a photographic plate, and some of the changes taking place in crystals and in solutions are noteworthy. Yet organic changes due to disturbance from neighbouring agents have

[1] In this statement, *Ether* is classed as a substance.

characters which distinguish them even from these.

Now it is true that there is no substance in inorganic Nature which approaches protoplasm, either in complexity of structure or in instability. Its elaborateness of composition is such that complete analysis is barely possible; and observation shows that it is in constant change, assimilation and waste being incessant in an endless conflict of loss and gain. The reactions of such a substance would be necessarily complex in character and difficult to trace. They may be wholly dissipated in internal changes of energy, which can, at the best, only be perceived indirectly and may escape observation entirely. There is some advantage in regarding the phenomenon of organic sensitiveness as a very complicated process, involving both chemical and physical changes in the material of the organism. But it would be a serious mistake to consider the final result of those changes, that is, the end of the process, as a direct product of mechanical, chemical, or physical causes. These give rise to changes in respect to space and time, *but sensation shows no possibility of measurement as a time-rate or as a space-rate.*

We are now, indeed, chiefly concerned with the end of the process, so far as there is any inference to be drawn from it with regard to organisms in general. We need to describe the final result of stimulus as exhibited by the organism as a whole. And in this connection it is found, that though some reaction usually follows a definite stimulus, it is not always of the same kind. The same stimulus may produce different effects in organisms of different structure. In addition to variety in effects, which may have a simple connection with difference in the structure of the organism upon which the stimulus acts, we have also to consider variation as due to the character of the stimulus itself. If we may argue from our own experience, differences in quality and intensity of sensation may be referred to differences in the nature and quantity of the stimulating change. There is, however, *no exact correspondence between the two groups of events as regards quantity, while the nature of the transmitting structure will determine the kind of sensation to be received.*

Finally, we may say that all reaction on the part of the organism to outer changes consists normally of a group of changes, varying according to circumstances, which have as their ultimate result the *advantage* and *preservation* of that organism. A description of the process in physical symbols would be impossible. It would be indeed strange if the organism were to impress itself upon us at this stage as a mere accumulation of quantities or substances.

Numerous difficulties have previously arisen,

needing the exercise of careful inference. But on the present occasion a demand is made upon a rarer mental faculty—the capacity to unite a number of generalisations under a still more comprehensive idea, and to ascend in the scale of abstraction. It is advisable that intellectual exercise of this character should be directed and controlled, for it is not always a gain to the student to initiate it. But here it is difficult to avoid borrowing from ideas gained in the observation of human activity, and we are bound to describe the essential character in the excitability of an organism as *subordination to a purpose*, not necessarily its own, and to a purpose which is common to all forms of life. The name to be given to that purpose is not so important as the recognition of its existence.

The Implication of Mechanism in Consciousness

In dealing with the event of sensation, as it may be studied in ourselves, we find a mechanism which is distinctly material taking part in the production of subjective results. A change of consciousness comes at the end of a material process. Recognising that consciousness never occurs, except in connection with material change (and full admission must be given to the fact), it is not surprising that the connection should be treated as causal. Now there is a common confusion between the end and the means of sensation, and in consequence the word *mechanism* is often considered inappropriate. The working of the machinery, if such a term may be used with regard to sensation, is antecedent to the change of consciousness. *It is in every respect external to consciousness.*

But it is no uncommon fault to use methods of inference, which have served adequately for one plane of thought, on a level at which they are useless. The mental momentum carries one beyond the proper boundaries. Most errors with regard to sensation arise, partly from a confusion between physical and psychical events, and partly from the undue extension to the world of consciousness of generalisations derived from material phenomena. The effort to find a general law for purely psychical events is legitimate, but it must be looked for elsewhere than among material or mechanical conceptions.

The tendency to erroneous conclusion is especially encouraged by the fact, that there is nothing in the organism which shows, more clearly than the organs of sensation, the distinctive requirements and form of a mechanism. But the organ or medium of sensation is not on that account to be confused with its end. Still, however uncertain may be the ultimate result of a given sensory change: whatever distinctions may need to be drawn between effects which are permanent and those which are transient: one thing is certain, *that the immediate result must depend upon, and be conditioned by, the apparatus which brought it about.*

And it is, moreover, worthy of notice, that the whole nervous system, as illustrated in the more distinct structure and the more differentiated events shown by the higher organism, gives a strong impression of an intricate mechanical system, to the interests and ends of which all the other activities are subservient. In any consideration of the organism's relation to its environment through the agency of its sensory apparatus, this view clearly suggests itself.

However persevering may be the effort to bring all natural science under the sway of general laws, there must be some failure, or knowledge would be complete and prediction infallible. And this is what is felt when we say that the region of Biology appears to be, in parts, *under the sway of chance.* Not only as regards the subordinate events within the organism, but as regards also the total relation to environment, there are many gaps to be filled in, before we can infer what will follow from a given set of conditions.

But in sensation we can always say what will *immediately* follow, and, as a matter of fact, the greater part of the process has been traced step by step in a series of physical events. The apparatus of sensation contributes to certain well-defined and observable changes, and most of their conditions are known. But these events are quite distinct from the *final* and *resultant changes* produced by them on the organism. Just as happens in the action of a machine, the sense-organs can do their allotted work, but they must do it in the fashion which suits their construction. They are unable to vary their methods, and no other course but that prescribed by their structure can be followed. The stages of this process take their form from the transmitting and intermediary organ. But after these occurrences have ceased, sensation begins.

A principle previously emphasised must be again repeated. The organism, as a whole, can and does decide for itself, whether further results shall follow the direct results of external change. And further results of the most varied character may occur. In a physical system each change produces its own results independent of others, but in the living system, through the exercise of a distinctive control, a variety of stimuli may be in operation, while one alone affects the organism. And this quality, which is called *attention* when we are dealing with sensation, is but another manifestation of that assertive character of adaptation, on the part of the whole organism, towards a distant end which is unknown to us, save as regards the fact that self-preservation is a motive in its attainment. Adaptation itself is an objective occurrence, it must be remembered, or these statements could not be made.

It has been an advantage to illustrate these points by referring to a more highly-organised animal. The difficulties are not so impressive in cases where variety of stimulation is unknown, and when nothing is inferable except a general susceptibility to environment. This is not the first occasion on which it has been found that organisation is an aid in making an examination of function.

The Sentient Subject

A still wider problem arises from a subjective aspect of the relation between the individual and its environment. Reflection on this question gives ample opportunity for the perception of a distinction between living and non-living matter, for here we touch upon the greatest of all distinctions — sensation as affecting the subject. But inasmuch as there is always a danger of confusing the objective facts of reaction to external changes, *i.e.*, the phenomena of excitability, with the subjective facts which are the accompaniments of that reaction (as far as our own experience of sensation gives information which is generally applicable) it is necessary to repeat, that in sensation we deal with a phase of life which has no parallel in any other mutual action.

More than once our attention has been drawn to the complete failure of every mental effort to bring together, for any purpose of comparison, things which are so entirely incomparable as the inner and outer processes of the organism. The difficulty presented itself at the beginning of our course, in the distinction which had to be drawn between the subject and the object of consciousness; and it gains further meaning from reflections which are appropriately given here.

The facts of the subject—the inner processes of sensation, thought, and will—are connected with the facts of the object—the phenomena of matter and motion. But to consider them as the same in kind would imply that movements of matter, at some stage or other, pass into still more subtle movements, which *are* colour, sound, or taste; and that if research has failed up to the present, yet at some future time the observer will be able to say: *Now is the moment at which matter or motion becomes sensation, and here is the mechanism by which the transmutation is effected.*

There is but one reply to be made, and it is an indirect one. The subject feels, it is itself the *cause of its feeling*. Living matter, or more accurately, the individual organism, alone seizes the outer events and fashions them to its own end. The form of the effect is determined by the organism. It is shaped by the nature of the organism, not by that of *its material ante-*

cedents. Each group of facts must be isolated, and the relations within the groups investigated separately. What is important for the student of Biology to recognise is the essential difference between the two groups. Further he cannot go, unless he wishes to enter on a research which is metaphysical.

The organism from this point may be looked upon as something *altogether different from what its material relations might be expected to make it.* And there is probably an infinite variety in this difference, in accordance with the variation in organisms. In the previous chapter we looked to the environment for some principle of development, and in that search learnt again how few biological theories are unconditionally true. And here we refer to the same limitation in saying that in one sense the organism creates its own environment. *All the same, there should not be any confusion between the objective effect of physical surroundings and the image of those surroundings as conceived by the rational individual.*

The term *individual* is meaningless, until a conception is gained of the subject receiving impressions from without. The idea of individuality, indeed, is overlain by that of the subject. To understand the individual, we must understand what is meant by singleness of relation between the subject and its environment, remembering that the boundaries of the environment are drawn by the sensibility of the subject. And we shall quickly perceive that difficulties will be encountered if we attempt to apply the same ruling to all organisms alike. The view of a relation between the living object and its external world, which implies the existence of the same kind of feeling as our own, can scarcely be extended to all organisms indiscriminately.

The utmost we are entitled to do is to infer for the lower forms of life a nebulous state of something resembling feeling. Whether we can extend to them anything corresponding with our own perception of variation in quality or intensity of sensation is also doubtful. To them, as to ourselves, objects which do not influence the sense-organs must appear non-existent. And we may, from morphological analogy, associate indistinctness of sensation with narrowness of range. There need be, then, no hesitation in ascribing to the lower organisms an incomplete individuality. The difficulties which arise in the application of the term individual to certain of the lower organisms are natural. With a better apprehension of the meaning of individual, the application would not be attempted.

CHAPTER XIV

SOME OF THE PROBLEMS PRESENTED BY THE ORGANISM

Recapitulatory

THE problems before the student of Biology are no less numerous than the generalisations which embody his normal experience of the subject. As soon as details are united together in the form of general statements, reflection becomes at once more active and inquiry more eager; and in the study of life, indeed, the step from generalisation to speculation is always especially short. The diversified appearances of life cannot fail to present exceptional opportunities for the exercise of inference, and for that intelligent guessing which reaches here and there to the dignity of recognition as theory. But it is a special feature of Biology that many of its conclusions cannot be put to the test of direct observation, nor are they always the simple instances of inference which they appear to be.

On the contrary, many of the questions which arise are only to be answered after a careful balance of conflicting evidence, and by the aid of an exact sense of the difference between realities and appearances; while many of the principles under discussion, since they involve conceptions which are not altogether physical, are beyond the reach of purely scientific judgment.

To ask questions is natural, but it is philosophic to recognise that there is a limit beyond which questions cease to be profitable. This limit is reached when answers cannot be given, or when all answers are equally meaningless. The existence of metabolism, specific reaction, and species are minor issues which make up between them the greater problem of life, and these serve as illustrations of questions which must remain unanswered. It is strange that the derivative problems should be considered more easy of solution than the larger abstraction upon which they are founded. Yet it is clear that a final answer to any one of these questions would involve an explanation of the existence of life itself, a task which is obviously impossible in a scientific sense. It may be reasonable to inquire *how* an organism grows, and there is plenty of information recorded thereon. But to attempt to learn *why* it grows is to struggle in vain against the barriers of reason. There is nothing here about which positive statements can be made, and certainly nothing to which the methods of observation apply. It may, in fact, be easy to ascertain how an organism reacts to external changes, while there may be no meaning in our asking why it reacts at all. Biology is a science which deals with processes and not with first causes. It cannot inform why there is life.

The manner of growth, or metabolism and its conditions: the extent of variation: the process and necessary form of reaction: and the systematic comparison of structures are legitimate subjects of biological inquiry. And though in their treatment a reference to the facts of Physics and Chemistry is a frequent necessity, the application of the latter to the facts of the organism does not constitute an explanation. Biology may need assistance from other sciences, but its chief problems continue, nevertheless, to form a class by themselves.

Species

The existence of the organism in different forms makes an early demand upon our aptitude for arrangement, but the establishment of order is soon followed by an effort to ascertain relations between the items of which the scheme of arrangement is composed. No sooner are differences clearly abstracted from the objects presenting them, than a distinction in differences becomes perceptible; and the perception of degrees and distinctions in unlikeness is only narrowly separated from the desire to bring all forms and all differences into the scope of a general plan. The disentanglement or analysis of diversity is usually followed by a synthesis or union of the components previously isolated. In the theories which aim at connecting the various parts of the scheme of classification we have an instance of the natural course of mental events.

Now a mere description of individual forms, even when the results of classification are appropriately employed therein, cannot be expected to convey the full meaning of its terms. The fact of species, indeed, together with what that fact implies, covers the whole ground of Biology. The recognition and arrangement of differences of form may be achieved by force of observation at the outset, though very different qualities are exercised in attempting to understand why the organism is complete and individual under its different forms. For ordinary purposes, differences between species are stated in the terms which are customary in all descriptions of phenomena, and in this connection the attributes of matter, namely, shape, size, mass, and kind, appear as the chief elements for consideration.

Nothing but the ordinary relations to matter and space which define every object are presented by the organism, though with regard to time both ordinary and extraordinary conditions hold. Time is the necessary medium of all events, whether they occur in inanimate nature or within the material boundaries of an organism. It is a conception abstracted from every kind of change, and it may be referred to them afresh. But in viewing the organism as a whole a further and distinctive condition as to time may be asserted. It is familiar to us in the conception of the duration of life in the individual. It is less familiar and more difficult to realise when applied to the species. For though the species may have a finiteness in time such as the individual shows, yet in attempting to resolve the condition of existence for either individual or species into its simplest elements, it must not be forgotten that time itself is a conception derived from our mental attitude towards events. We do not approach necessarily any nearer to ultimate realities by translating appearances of this kind into terms of time, space, and matter; nor, on the other hand, can it be asserted that the main problem of the existence of species is rendered more difficult by the recognition that specific forms are not stable.

It is this aspect of species which has given rise to several important conceptions, notably, *evolution* and *natural selection*. In evolution we have a hypothesis which assumes that both the changes and the forms which present them progress spontaneously from simple to complex. In natural selection an agent is provided, by means of which a progressive operation manifests itself discontinuously and species are brought into existence. Other relations than these have been suggested as existing between divergent forms, but the scope of our inquiry does not include such topics. It is enough for our present purposes to draw attention to the more elementary conceptions which ought to be grasped before speculation begins. And chief among these conceptions is that in which variety of form is mainly regarded as the more immediate manifestation of variety in modes of activity. Yet it would be a mistake to consider that the kinds of activity are fundamentally different. It is rather a difference in the proportions in which their constituents are combined that gives variety to the dynamic aggregates. Variation ensues, it is evident, more in the mingling of component activities than in a fundamental difference of those activities. The material and the form may be less significant than the agent, yet though we may translate the question of form into the language of change, transferring our attention from the medium to the event, we are still without an answer. Organs and structure, even if only the expression of dynamic differences, are parts of the greater problem of life.

We gain little in this connection from the perception of the adventitious nature of much of the presentation of form, or, in other words, that much of the material giving the organism its spatial relations is often, and notably in vegetation, the refuse of previous activity, and to that extent external to the organism, the excreta becoming in this case a vehicle of extended activity. It is clear that such material might often be replaced by inorganic substitutes of similar mechanical functions without injury to the organism. Ideas of this kind are reproduced in the modern conception of the cell, in which the difficulties of form are repeated in miniature. The protoplasm of the cell is in this light no longer the primitive material of life. It is found to be too complex in structure to serve that purpose, and its characteristic events are traced to a still more fundamental plasmogen within a less active medium.

The existence of species is, in reality, the same class of phenomenon as the isolation of the individual. Discontinuity of life as regards space and time resembles very closely the divergences in details of structure. It is difficult to understand either individuals or species, except as products of the same order; and this view becomes clearer and more consistent, if we endeavour to regard both individuals and species as occasions of change rather than substances.

The conception of form is extended in application by those important ideas which are represented by the terms *type* and *homology*. A repetition of parts and an indication of pattern is a character of all but the simplest organisms. The breaking-up or discontinuity of processes is evident in the distinctness of types, while the similarity of the agent is obvious in the existence of homologies. In assuming the existence of an agent behind the facts of form, we are doing no more than is done under other circumstances in using the term gravitation. Both modes of expression are convenient, inasmuch as they bring about an economy of language. From one group of material changes we abstract the general term *gravitation*; from another group of facts we abstract the idea *life*. To say that the agent, life, makes itself known in discontinuous individuals, species, or genera is to re-state the fact, that from observation of such discontinuous things existing in Nature we have formed a conception or abstraction called Life.

This, then, is the agent to which reference has been made, and it is apparent in parts as well as wholes. Its fragmentary and limited manifestation, which is described by the word discontinuous, is even more strikingly prominent in the homologous disposition of parts. For if the operation of life be not always discontinuous, and, in examples of the same type, similarly

S

discontinuous, how are we to regard the preservation of the same relations of parts to one another which is evident in many different species? In all those individuals which belong to the same type, we have structures exhibiting a similarity of pattern or design, wherein appears a manifestation of similar modes of activity. We perceive in them a discontinuity of the same kind, and the evidence of a succession of events proceeding along confined paths in accordance with a rigid scheme. The maintenance of similar structures even in different species may therefore be regarded as a clear indication of limitations in the objective manifestations of life. Life itself is continuous everywhere, but everywhere its vehicle is checked and its material presentation is restricted.

Variation and Adaptation

The perception of homology has, indeed, proved a most valuable factor in the development of biological ideas. Its influence has been widespread. A marked advance has followed on the comprehension of similar objects occupying correspondent positions in the configuration of distinct specific forms. It has now become easy to realise that such correspondence, or homology of material structure, corresponds with similarities in a uniform scheme of activity. In different individuals we may see indications, not so much of the similar results of independent efforts, as of the same agent selecting within definite limits the same medium for its manifestation.

It is always well to remember that the highest point to which observational methods can finally attain is the discovery of further relations, and given the existence of such facts as are evident in all species of the same type, the suggestion of a connection by way of descent is a step in advance. Yet the progress here is not great, for the relation is no better known than the facts it connects. A new form of expression incorporates the old impressions without disposing of them. The relation of parent and offspring, which is thereby extended in scope, is no easier to analyse than the relation between dissimilar forms. The genesis of an individual and of a species must be regarded alike as ultimate facts of Nature.

In either case, an agent external to the event needs to be assumed, and those suggested have each in turn given rise to much controversy. The phenomenon of repetition of the individual involves the assumption of heredity; and the co-operating factors in the origin of species by descent are usually taken to be variation and natural selection. A little reflection, however, will show that both terms, heredity and variation, cover the same gap in our knowledge. While we are ignorant of so many details, and, in fact, of every detail which has a quantitative or positive value in the phenomenon of descent, even in its most immediate instance, abstract terms of this vague character are a necessary part of our descriptions. The relation as regards matter in the case of descent is enough to force the admission, that the main occurrence is the inheritance of a tendency or a mode of activity. Yet any attempt to define the meaning of the word *tendency* in this connection will probably modify any feelings of certainty which we may happen to have about our knowledge of these events, and it will perhaps be admitted, that our difficulties are not removed by an estimate of the completeness or incompleteness in the descent of the tendency implied in heredity and variation.

Nevertheless, it is certain that the study of variation can give valuable information as to the processes of life. But however successful in detail that study may be, and however attractive it may be to speculate on these matters, it is well to remember that variation, heredity, adaptation, and selection are mutually inclusive ideas, which can scarcely impress the inquirer as having the stamp of finality.

In dealing with these ideas it is important also to recognise that the same problem lies before us, whether we look for its solution in the organism or the environment. There is no purpose served in attempting to explain one side of a dual manifestation in terms derived from the appearance of the same event from another side. This caution is specially needed when the relation of the organism to its environment, which is the main feature in questions of adaptation, distribution, or selection, comes under discussion. Whether, for example, degree of adaptation is a true measure of the extent of life; whether the cause of species is an active agent within the form itself, or the form a result of passive reaction to environment; by what means a continuous action corresponds with a discontinuous reaction,—these and similar questions are best discussed when the larger difficulty is appreciated. It is indeed probable that they may be all stated afresh in the single question, Does the form unfold itself from pre-existent forms, or is the form created, *i.e.*, does it " become," or is it always in existence?

Problems of the same order must arise in connection with the generalisations of embryology. The course of development, with its apparent record of ancestral structures, may be looked on, it is true, as an indication of an origin by descent. On the other hand, it may appear no more reasonable to see in these facts the evidence of a common origin than the result of a constant environment operating on an organism, which reacts and adapts itself, in diverse ways, it is true, but with limitations which are obvious to all who consider the restricted number of types or the fewness of functions in existence. Nevertheless, whenever the action of the environment is put forward as an agent, either of variation itself or of limitation in variation, it is helpful to remember that two germ-cells may give rise while in the same environment to widely different adult individuals, even though the inherited embryonic structure, which is assumed for this effect, is regarded as due to the influence of environment in the past.

The Cell as Agent

The universal evidence of a cellular structure in organic objects has not brought about exactly the results which were expected. The cell has proved to be a morphological guide. It has enabled us to form an image of the process of development, and the mere dissection of any biological process into its details is a gain. But the assignment of distinct and definable stages to the process which is most marked in the life of the organism has not rendered the process easier of comprehension. It has undoubtedly reduced the structural whole to its real component parts, and it is important to recognise that the most valuable result of this analysis lies in the opportunity which it affords of comparing different structures with one another. By reason of similar structural elements existing in different composite wholes, a comparison in terms of the elements becomes easy and instructive. Differences in structure can now be referred to variations in the mode of arrangement of similar parts, and the perception of similar parts in different products facilitates generalisation.

But the conception of the individual organism is not rendered more simple. On the contrary, it gains in complexity. The various modes of cellular combination may take the place of difference of species, but the old difficulties reappear. The question remaining to be answered has been changed in form without being solved. The apparent adaptation of the cells to their end does not explain the adaptation of the aggregate to its end. Variation in the form of kindred cells does not show why similar functions should be localised in widely different aggregate organisms. Indeed, an additional problem presents itself. When it is remembered that the reality underneath structure is a metabolic process, that it is more a question of modes of activity than of form which has to be settled, it will be seen that the adaptation of each subordinate activity to its place in a consistent scheme is a fresh problem of considerable difficulty. The primary facts of the organism which are denoted by the terms specific difference, variation, heredity, and others, are repeated in the cell, while the further difficulty of the origin of the co-ordination of cellular changes is created. The consensus of a group of minor activities to the purpose of a larger change is one of the chief forms in which the many-sided problem of life can be stated. It would appear, indeed, from consideration of unicellular organisms and of the germ-cell itself, that in the complex organism we have a group of more or less subordinate organisms, more subordinate in the light of their relation to the whole, and less subordinate so far as they exhibit in themselves the attributes of the whole.

The Unity of the Organism

But the agency of the cell is not the last stage which has been reached in the course of research, nor is it likely that the process of analysis will ever reach its limit in localising the activity of the organism. A closer and closer confinement of the formative activity, first to the cell, then to the nucleus, and later to special parts of the nucleus, points persistently in that direction. The admission that protoplasm is not homogeneous, but an aggregate of various components, is only one more step in the search for that unrevealed substance which will always remain behind that which is the last disclosed.

The least elusive fact is that of change; the dynamic of life is less deceptive than the form. A perpetual redistribution of matter is a universal accompaniment of what we perceive to be life, but the main tendency of this change, whether, as implied in evolution, it is progressive or not, can only be decided by reference to the individual organism in connection with its surroundings. It is possible to regard the organism as the sum of its activities, but better to look on every organism as the outcome of a single operation necessarily appearing to us in fragments. The addition of all its chemical and physical changes does not yield the organism as a total. To attain to an adequate conception of the unity of the organism it is not enough to bring together elements which have been derived, in the first place, from an analysis of that unity. It would be an advantage, therefore, were our positive knowledge adequate for such a plan, to start the course of our thoughts from life as a whole, and trace out as completely as possible the varied manifestations of an universal activity.

The singleness and completeness of the individual is the end no less than the beginning of biological inquiry. The apposition of the isolated organic experience to a multiplicity of objects, which appear to serve as the occasion of that experience, is a mode in the working of our intelligence which reasserts itself in the necessity under which we labour of referring every change and manifestation of life to the completeness and individuality of the organism. It is difficult to formulate a statement, clear enough to carry conviction, as to the organism in its environment being a dual manifestation of a single reality; and the duality of the world, expressed at the outset in terms of subject and object, and repeated in the more precise conception of the isolated individual reacting to its environment, cannot be brought together under any single image which would be in keeping with the scope of these pages. At this point, in fact, our inquiry, would lead through the path of philosophy to the region of beliefs. The undisclosed residuum remaining when all physical phenomena are abstracted from the individual is very naturally referred to an agent or cause of unity. Every branch of research leads sooner or later to the unknowable, and that stage is soon reached in attempting to pursue the mechanical conception of life.

INDEX

ACTION, automatic, 173
  illustration of mutual, 231
Activity, composite nature of, 186
Adaptation, 258
  as measure of life, 261
Alimentary canal of man, 30
Alimentation, structures subservient to, 43
Analysis, value of, 184
Animals, relations with plants, 131
Assimilation, 87
Attention, 109
Authority, 72
  an index to point science has reached, 74
  necessity in education, 73

BIOLOGY, definition of, 10
  difficulties of study of, 11
  embraces two classes of ideas, 20
  includes feelings and appearances, 16
  province of, 112
  recapitulation of problems of, 250
  scope of, 27
  unsolved problems in, 154

CELL, as agent, 262

Cell, reservations in use of idea of, 204
Cell-arrangement, importance of idea of, 219
Cell-division, accompanied by differential growth, 208
  differences between in embryo and adult, 207
  incident of growth and reproduction, 200
  unlike products of, 206
Cell-theory, 205
Change, 21
  characteristic of life, 196
  kinds of, in living things, 163
Changes of matter common to all forms, 25
Chemical changes, 137
  in relation to our consciousness, 139
Chemical methods, limitations of, 128
Chemical nomenclature, 130
  does not aid in explaining life, 140
Chemical principles not inclusive in biology, 142
Chemistry, principal ideas of, 150
Circulation, results of, 39
Circulatory system, 34
Classification, 56

268   THE LIVING ORGANISM

Classification, based on groups of characters, 70
  by generalisation and abstraction, 58
  development in, 65
  independent of ultimate cause and origin, 72
  must have some central idea, 61
  must regard both form and function, 62
  not based on superficial characters, 60
  not necessarily a simplification of outlook on nature, 59
  order the first object of, 57
  perception of distinctions in, 65
  perception of similarity, beginning of, 58
  scientific system of, 63
  summary, 76
Cohesion, effect on organism, 192
Combination of functions, 219
Conservation of matter, 138
  of energy, 164, 233
Consciousness and mechanism, 242

DEATH, meaning of, 84
  association with reproduction, 90
  sequel to reproduction, 84
  sequel to reproduction for the cell, 202
Degree, meaning of, when applied to form, 177
Development, meaning of, 66
  defined in relation to form and change, 68
  nervous, 221
  two paths of, 215
Differences between animals and plants, 21
Diversity of species, cause of, 223

ELEMENTARY substances, 138
Energy, conservation of, 164, 233
  doctrine of, in relation to biology, 161
  forms of, 164
  no explanation of growth and development, 174
  of animals, 166
  of living things derived from sunlight, 170
  of plants, 166
Environment, attention to, important, 212
  cause of variation, 260
  changes in, 213
  extension of idea of, 214
Evolution, 209, 254
Excretory system, 36

FEEDING, characteristic of living things, 53
Form, meaning of, 7, 21
  conception of, 181
  extension of idea of, 257
  externals of life, 180
  intermittance of, 82
  number of varieties indefinite, 6
  permanence of, 91
Function, localisation of, 187

GENERALISATION, 22
  limit of, 26
  process of, 126
  relates to aggregates, 24
  simplifies description, 24
Gravitation, effect of, 191
Growth, 85, 203
  assimilation underlying, 87
  connection with reproduction, 88
  relation to cell division, 204

HOMOLOGY, extension of conception of, 257

INDEX 269

Homology, perception of, 258
Hypotheses, utility of, 217

IDEAS, definition of, 111
and sensation, 111
Individual, as subject, 248
division of, 95
idea of, 94
identity of, 88
problem of genesis of, 259

KNOWLEDGE, and language, 74
based on perception of similarity, 149
derived from observation, 8
grows exact by communication, 74
nature of, 4
of material objects, 148
of matter, 145

LANGUAGE increases determinateness of thought, 74
Life, abstraction of biology, 257
choice characteristic of, 55
continuity of, 81
dependence on food of, 54
equilibrium checked by, 224
every act of, a change in matter, 119
lack of movement in plants, 51
material basis of, 118
movement characteristic of, 51
utilisation of processes characteristic of, 143
variety of forms of, 178
Light, influence of, on organism, 192
Living material, characters of, 121
relation to parts of an organism, 157

Living system, the, 190

MATERIAL basis of life, degree of resemblance in, 123
difficulty of perceiving as common, 122
Matter, description of, 143
knowledge of, 145
Mechanical analysis, 37
Metabolism, 141
Morphology, 182
distinguished from physiology, 64

NAMES of elements, 134
Natural science deals with phenomena, 104
Natural selection, 255

OBJECT, constancy of, 109
relation of to subject, 98
Observation, beginning of exact knowledge, 5
control of, 127
engages mind, 103
restriction of, 152
Organic activity, difference from that of other systems, 173
resemblance to that of other systems, 170
Organisation of anthropoda, 47
of birds and mammals, 45
of echinodermata, 47
of fish and amphibians, 45
of lowest animals, 48
of molluscs, 46
Organism, activity of, 179
aspects of, 61
both subject and object 105
contrasted with mechanical system, 141
creates its own environment, 248
effect of external changes on, 191

Organism, energy of, 161
  gain of energy in, 195
  inseparable from surroundings, 106
  meaning of, 62
  not an ordinary conservative system, 168
  reaction of, 193
  unity of, 264
  viewed as an aggregate of certain kinds of matter, 140
  with environment to be treated as a system, 188

PERCEPTION and sensation, 111
  aspects of, 13
  basis of knowledge, 2
  definition of, 111
  direction needed, 4
  implies life, 14
  implies perceiver, 13
  of similarity involves classification, 58
  range of, limited, 3
Physical changes, classes of, 162
Physical methods applied to biology, 159
Physical principles, ideas of, 150
Plants, food of, 49
  relations with animals, 131
Protoplasm, 153
Pseudo-explanations of growth and development, 197

REACTION, 193
Reproduction, association with death, 90
  conditions of, 88
  connection with growth, 88
  discontinuous growth, 83
  repetition of forms, 82

Respiratory organs, 33

SCIENCE does not treat of things in themselves, 147
Sciences, relation to one another, 14
  sources of error in, 15
Self, ideas of, 110
Sensation beginning of thought, 107
  cannot be measured, 240
  change in matter antecedent of, 234
  character of, 238
  consideration of, 107
  in relation to changes, 236
  link between animate and inanimate matter, 237
  link between organism and environment, 116
  meaning of, 229
  reality of, 146
  result of, 241
Similarity of substance, 124
  test of, 151
Species, 252
  origin of, 254
Stimuli, 235
Structure, cause of diversities of, 182
Subject, assertions concerning, 101
  ideas of, 100
  relation to object, 98
  relation to sensation 246
Substances limited in number, 6
Summaries, classification, 77
  feeding, 51
  form and development, 225
  individual, 92

TEMPERATURE, effect of on organism, 192
Thought implies more than sensation, 108

Types, conception of, 69
  recognition of, 71
  serve as standards of reference, 70
UNITY under diversity of form, 63

VARIATION, 258
  results from mingling of components, 255
  self-originated, 213
  variety of form, 155
WORK, 165

THE END

Printed by R. & R. CLARK, LIMITED, *Edinburgh*

www.ingramcontent.com/pod-product-compliance
Lightning Source LLC
Chambersburg PA
CBHW032120230426
43672CB00009B/1801